照明设计的故事
Lighting Design Stories

英国莱亭迪赛灯光设计合作者事务所 - 中国分部 编著
Lighting Design Partnership International (LDPi) China Branch

2002-2014

序

照明设计在中国

2002年，在漫天飞雪的冬天，我从爱丁堡来到北京。没想到，从北京香格里拉酒店项目一直做到今天。照明设计这十年在中国也蓬勃地发展起来。

我是在英国伦敦大学学习比较文学之后，又学习设计管理。学完之后，阴差阳错进了爱丁堡的LDPi（Lighting Design Partnership International）开始学习照明设计，一边学习一边从业至今。

过去十年，中国经历了房地产行业的快速成长期，住宅和综合楼开发如雨后春笋。作为早期一批海归之一，在市场化大背景下我得到了很多案子。我见证了照明设计在城市化过程中，从无到有发展的十年。

在中国刚起步时，LDPi接触到的更多是住宅和酒店方面的项目。近几年，我们赶上了一个商业地产的开发时代。展望未来十年，互联网时代的到来，商业环境会更加重视客户体验，重视独特化和差异化的空间展示风格。我想，我们会迎来一个更好的、更加需要照明设计的时代。

除此之外，在设计理念上，我们设计的案子大部分不追求太过时尚的方案，保持了安静对称和韵律感。时尚的另一面很容易走向世俗，我们坚持自己是建筑照明设计公司，而不是舞美灯光设计公司；我们坚持用持久一点的眼光去看大千世界，而不是追求一瞬间的眼球效应。正是这种坚持，让我们赢得了万科、香格里拉、悦榕庄、凯德置地、万豪、希尔顿等这样一些以筑就人文建筑为理念的开发商和酒店集团客户。

不同于其他规划院，我们很少接触政府的项目。一方面是因为政府的项目一般要求本土的一些资源，另一方面是因为政府很少将照明设计和工程分开做。因此，很多城市项目出现用灯过度的现象。我们也很少做一些规划项目，因为很多规划最终很难保证它的完成效果。

这十年，我要感谢每一个与我们合作过的客户。其实，很多客户对照明认知不足，预算的紧张，让灯具产品价格在一定程度上成了悬在设计师头上的达摩克利斯之剑，使得设计师只能更加努力去为客户选择合适的灯，而不是贵的灯。一个好的设计师就像一个好的乐队指挥，他把不同的乐器组合在一起，并为客户创造完美的一首曲子。我们坚持用这种方式回报客户。

每个客户对于自己项目的了解是非常深刻的。我们在介入项目后，总是能够从客户的案子里，发现很多优秀的合作伙伴，包括建筑师、幕墙顾问、景观设计公司和室内设计公司等。记得有一次与英国"设计周"创始人约翰·佐雷尔（John Sorrell）聊天，问起为什么英国会产生那么多优秀的设计，他回答说，那是因为英国有很多优秀的业主，对此我感受特别深刻。如果没有凯德置地开发商对北京来福士项目外立面照明方案的力挺，可能今天在北京东直门交通枢纽的来福士夜景不会是这个样子。他们坚持让结构方（Arup）设计一款可以从外面检修的窗户。这样，虽然在设计之初麻烦一些，但避免了在建筑室内摆放灯具的方式，同时保证了立面的灯光效果。

我还要感谢一些在国际上设计了很多了不起建筑的建筑设计公司。记得当时有一个客户让美国斯蒂文·霍尔建筑设计事务所（Steven Holl Architects）来面试我们，在沟通时李虎先生问："作为一个照明设计师，您觉得最难的部分是什么？"我说是对整个流程从始至终的专业把控，它需要做大量的工作。他说："是沟通吗？"我说是。其实，每一个专业设计师的成长都离不开

Preface

Lighting Design in China

In a snowy winter of 2002, I came to Beijing from Edinburgh. Little did I imagine, from the Beijing Shangri-la Hotel project to today's projects, I have witnessed the rapid development of lighting design in China in the past decade.

After graduating from the University of London SOAS with a degree in Comparative Literature, I went to study Design Management. Coincidently, I joined Lighting Design Partnership International (LDPi) in Edinburgh to work in lighting design and have started my career in this field since then.

The real estate industry in China has experienced a vigorous growth in the last ten years. Residential buildings and complexes sprang up rapidly. As one of the early overseas returnees, I have signed many projects in the Chinese market and have the fortune to witness the growth of the urban lighting design.

At LDPi's early stage in China, residence and hotel projects formed a major part of the company's profile. The commercial property is blossoming in recent years. In the next decade, with the beginning of Internet era, customer experience, individualized and differentiated space design will be much more important in commercial environment. An era with lighting design becoming even more necessary is coming.

Nevertheless, fashionable design is not always the pursuit of LDPi. For most of the time, our designs are rich in quietness, symmetry and rhyme. Being too fashionable more or less plays with the danger of being vulgar. We believe that LDPi is an architectural lighting design company rather than a stage lighting design company. We would like to look at the world in a more perdurable and sustainable way. Temporarily catching eyeballs is not what LDPi is here for. This very concept gains us many clients who are more cultural oriented, such as Vanke, Shangri-la, Banyan Tree, CapitaLand, Marriott, Hilton and etc.

As opposed to planning institutes, LDPi seldom works on government project. One of the reasons is that governmental projects often require local resources; another is that those projects rarely separate lighting design and construction. Therefore, the abuse of lights has become a common project problem in many cities. Additionally, planning projects are also rare in LDPi's profile since the final results of many plannings are hard to guarantee.

I would like to express my gratitude to every client that has cooperated with LDPi in the past years. Clients' unfamiliarity with lighting and the limited budget often create challenges to lighting designers. Therefore, we always work even harder to select the most appropriate lighting device rather than the expensive ones. A good designer is just like a conductor of an orchestra, gathering different kinds of musical instrument that performers harmoniously use to play a beautiful melody. In our endeavors in the past decade, LDPi cares most about how to help its customers achieve the highest return on investment.

The clients are the true master of their own projects. Each time when our design team joined a project, they have always been able to discover many outstanding working partners, including architects, curtain wall consultants, landscape design companies, interior design companies and etc. I remembered once I talked to John Sorrell, the founder of UK Design Week,

良好的沟通技能。照明设计师面对的沟通可能更多，假定一个照明设计师接到的是一个从里到外的设计项目，他需要紧密沟通的方面如下图所示：

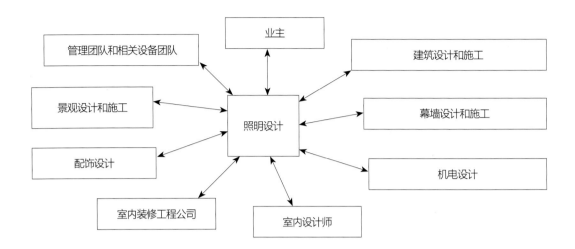

所以，一个设计师除了具备良好的专业设计水平之外，把握和谁沟通、何时沟通的能力也非常重要。客户并不一定知道什么时候轮到你上场，所以一定要积极告诉客户，哪些时候照明设计师是必须要去配合的，并且应该明白自己到底需要预留250毫米的灯槽，还是小一点也可以。

我还要感谢所有坚持与设计师共同实现完美效果的工程承包商和厂商。然而与此同时，我们坚持独立于厂商和施工方之外。这种坚持，赢得了同行和客户的认可和尊重。这也是能够让LDPi作为品牌设计公司最基础的原则之一。

我也特别感谢LDPi的同仁们。其实，刚到国内时，很多客户不理解照明设计的工作内容究竟是什么，我们坚持设计加顾问的全案跟踪方式，以保证每一个项目的品质和实现效果。有时，甚至在项目完工后，还追着客户去调试一些已经调过但还不够完美的地方。这种坚持保持到今天，成了LDPi工作的惯例。

眼前的这本书，不仅是一个照明设计公司十年在中国做照明案例的呈现，也是对中国照明这十年历程的一个回顾。灯光，点亮生活中的细节，同时，也点亮每个人心中的希望。 我们赶上了一个好的时代，我们因作为这个时代的建筑照明设计者而觉得幸福。深深感谢每一个人！

丁平

asking why there are so many excellent designs in Britain. His answer was because there are many outstanding clients, which I deeply agreed with. Without CapitaLand's strong support in the facade lighting design of the Beijing Raffles City, there would not be such a magnificent night view from Dongzhimen. CapitaLand's insistence on the engineering consultant, Arup, with designing a window that could be maintained from the outside was troublesome at first. But eventually it was made possible to avoid maintaining lighting fixtures within the architecture and guaranteed the lighting effect on facades.

My gratitude also goes to those top architectural design companies in the world that create so many amazing architectures. Once a client invited Steven Holls Architects to interview us and I was asked by Mr. Li Hu as a lighting designer, which part I thought was the most challenging. I said "it is the handling of the whole process in a professional way as it needs a huge amount of work". He then asked whether it is because of communication. I agreed. A professional designer cannot live without great communication skills. Lighting designers may encounter even more communications in the course of a project. For example, if a lighting designer gets a project to be responsible for its interior and outdoor lighting, then the lighting designer needs to closely communicate with the following parties:

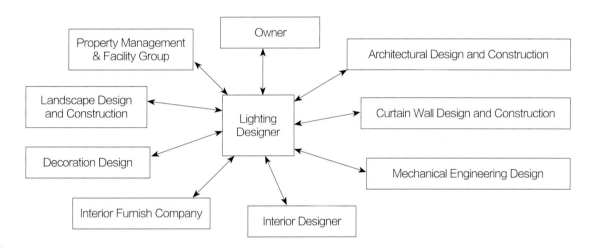

Therefore, a designer not only needs professional knowledge, but also needs to know with whom, when and how to communicate. Large chances are that clients do not know when they need a lighting designer to be involved, so it is of great importance to actively talk to them and let them know at exactly which point that a lighting designer is a must. Meanwhile, the lighting designer also has to clearly know whether a 250mm light cove or a smaller space needs to be designed.

I would like to thank each and every engineering contractor and manufacturer who devoted themselves to realize the wonderful lighting effects together with our designers. But at the same time, LDPi always maintains its distance and independence from manufacturers and contractors. Such persistence is widely accepted and respected by our fellow colleagues and clients, which is also one of the most fundamental core values that makes LDPi what it is.

I would also like to thank each and every member of LDPi. In fact, when I first went back to China, many clients did not understand what exactly a lighting designer does. Therefore, we insisted on providing a full-design consultancy to ensure the quality of each project and the final lighting results. Sometimes, after a project is completed, we even continue to work with our clients to resolve any outstanding point. This now has become a routine of LDPi.

This book not only presents a selection of cases of the past ten years LDPi has done in China, covering hotels, commercial complexes, office buildings, residential areas as well as educational projects. It also reflects the course of development of lighting design in China. Light, illuminates our life and hearts and brings us hope. We are lucky enough to be in such a good era and we are so happy and proud to be the lighting designers who help to switch on people's life. A million thanks to everyone!

Tewa Srilaklang

序

照亮世代

对于大多数生物来说,光是生命旅程中第一个也是最后一个接触到的临界物。在人生的漫漫长路中,因为有了光的存在,我们才可以交流、生存、见证美丽彩虹的形成乃至一个国家的变革。

据新华社 2011 年 1 月 7 日报道,2010 年中国的家居照明和商业照明约为 5 028 亿千瓦,占全国电耗的 12%。源自美国能源信息管理局的数据显示,2011 年美国的家居照明和商业照明约为 4 610 亿千瓦,约占全国电耗的 12%。

照明设计师带着自己的责任和环保意识加入到项目中去,与建筑设计师、室内设计师和景观设计师精诚合作,为中国的发展建设贡献力量。

在与客户的交流沟通中,照明设计师总是在建筑、室内、景观设计师的后一步加入到项目中去。从流程上来说,照明设计师与建筑、室内和景观设计师合作,并在他们依次完成设计工作并得到客户的确认之后才会介入到项目中展开设计。从历史角度来说,自新千年伊始,照明设计在中国进入发展阶段。如果将建筑、室内和景观设计比作为早已大学毕业并已步入职场的大哥大姐,那么照明设计则好似一个尚未高中毕业但已然显出良好潜力的出色少年。

LDPi 并不依赖于任何照明厂商或第三方,仅靠自己的设计质量和尽职尽责的团队将项目从头到尾完成。与此同时,LDPi 深知只有孜孜不倦地学习才能进步。于是,自 2008 年开始,加入了一个名为"TNT 照明侦探团"的非营利组织,致力于用实用的方法对照明文化进行研究。

社会责任感是 LDPi 建设的一部分。自 2009 年开始,其将利润的 2% 捐献给陕西省留坝县的孩子们,资助他们去上海长颈鹿美语学校参加英语夏令营。2010 年,LDPi 首次创办了 Switch on Beijing "创意点亮北京"国际灯光艺术节,现发展成 Switch on China "创意点亮中国",使更多的人有机会接触和了解创意设计,并享受照明所带来的视觉盛宴。

从本书所囊括的照明故事中,您不仅能了解到 LDPi 自 2002 年开始为中国的照明设计发展所做出的努力和贡献,同时您也能感受到灯光是如何通过团队协作改变观念的。衷心地希望本书可以使照明设计在中国就像在其他国家一样受到更多关注和认可。

祝 LDPi 生日快乐!

叶伟德

Preface

Lighting for Generations

Like most living creatures, lighting is the first and the last element that one experiences. But during the course of a lifetime, lighting contributes to witnessing, interacting and living in an environment, whether it is the natural formation of a beautiful rainbow or the changes of a country.

In 2010, residential and commercial lighting contributed 12% or about 502.8 billion kWh of the total China electricity consumption (Xinhua news 2011-01-17). In 2011, residential and commercial lighting contributed about 461 billion kWh, or 12% of the total US electricity consumption (US Energy Information Administration estimates).

Together with architects, interior and landscape designers, lighting designers have the unique opportunity to cooperate on projects contributing to the growth of a country such as China, certainly by keeping in mind that it should also be done in a responsible and sustainable manner.

When we meet customers, lighting design is always presented as a discipline which is one step behind architectural, interior and landscape designs. From a process point of view, lighting designers cooperate and wait until each design stage of architects, interior and landscape designers is completed and confirmed by the customer before starting its own lighting design stage. From a historical point of view, since the beginning of the new millennium, lighting design is in its secondary stage in China. It is like an outstanding teenager who has yet to graduate from high school, while architects, interior and landscape designs are recognized big brothers and sisters who have long graduated from university and are already in their professional career.

Being independent from any lighting manufacturer or third party ensures that LDPi relies solely on the quality of its design and the integrity of its team to manage a project internally from A-to-Z. At the same time, LDPi values learning as a way of improvement. Therefore, from 2008, it has joined the Transnational Tanteidan, a non-profit group dedicated to the study of lighting culture through practical methods, mainly by engaging in fieldwork.

Form 2009, as part of its corporate social responsibilities, 2% of its net profit is allocated for kids from the Liuba, Shaanxi province to attend a one month English summer camp at the Shanghai Giraffe English School; in 2010, it initiated the Switch on Beijing international lighting festival, making creative design with lighting an understandable and accessible visual pleasure for the mass, transferring itself into Switch on China.

The lighting stories included in this book not only reflect the fact that LDPi has been engaged in China's lighting design since 2002, but they also are stories about how lighting, with team work, can change perception. This book has the hope that the field of lighting design receives further acknowledgement and recognition in China as it already did in some other countries.

Happy birthday LDPi!

Dominique Phommahaxay

序

成长感悟

2004 年 3 月加入 LDPi 团队，经历了人生成长的重要十年。当时国内专业照明设计单位屈指可数，在一个朋友的介绍中了解到英国的一家专业灯光设计公司入驻北京，通过毛遂自荐非常幸运地得到英国总部 Lawrie 的肯定，受邀加入到 LDPi 团队中。参与过多个项目的设计和管理工作，如酒店、办公场所、住宅、学校、城市建筑综合体等，从项目中获取了照明设计经验，从设计管理中得到丁总的指导，从中西文化理念实践中得到了锻炼成长，以至于今天成长为公司的合伙人。

LDPi 打造了很多创新照明设计平台，如"创意点亮中国"、"TNT 照明侦探团"等很好的公益性活动，并得到了业内人士的高度肯定。在日复一日的工作中，照明设计是在任务的条条框框下发挥的，很容易被带入一个疲倦期。这让我想起了在做"创意点亮北京"活动时我的第一个作品"漂浮的迷宫"，当时我表达的是一种心态：人在迷宫里寻找出口的穿梭心境亦如此，如同我们新一代 80 后的照明设计师一样，在大城市拼搏中寻找自己的归属感。通向成功的出口总会在你坚持不懈的努力中寻找到。保持对照明设计浓厚的兴趣，尝试不同空间的照明设计手法，在设计成长中从考虑光线的构成到光自然状态的表达，每一次都是对设计的提高和跨越。现在我更多的是静下心来，关注自然光线带给我们的兴奋、舒适、暗淡等不同感受，希望把这种感受带到设计中去。关注光的舒适度，寻找光与空间的融合，而不是让人感觉光是附加的东西。光与人之间存在着一种互动和交流，唯有理解它，才能掌握和运用它。

感谢丁平女士！感谢 LDPi 团队！

用心感受是最美好的回忆。

张静

Preface

Apperception from Growth

I joined LDPi in March 2004 and the last ten years with the team have been the most important years in my life. There were still very limited lighting design companies in China, I learnt from a friend of mine that a professional British lighting design company entered Beijing and was pretty determined to self-recommend. Luckily, with the approval of Lawrie from the UK headquarter, I became a member of LDPi. After participating in the design and management of various projects, including hotels, office buildings, residential area, schools and urban complexes, I have gained extensive experience in lighting design and built up project management skills with our Executive Director Tewa Srilaklang's guidance. The continuous practice of the theories from Eastern and Western cultures also forged me to become one of the partners of the company.

LDPi has contributed to many platforms combining creative lighting design, such as non-profit events like Switch on China, TNT Lighting Detectives, which are widely recognized within the field. Working under various rules and requirements is a routine in lighting designers' career, which we may easily get tired of. My first artwork for Switch on Beijing is the Floating Labyrinth, expressing the state of mind that people may have while finding the exit in this labyrinth, which is just like those of lighting designers born in 1980s looking for the sense of belonging in a big city. The exit will eventually be found with unremitting effort. I told myself to maintain enthusiasm and interests in lighting design, to explore and try lighting methods in various spaces. Gradually, I began to think about presenting light naturally rather than merely considering the composition of lights. As the skills built up and the understanding towards lighting design enhanced, I now focus more on the feelings brought by natural lights to people and attempt to embrace them into my design; meanwhile I also consider more about the comfort that lights bring, seek for the perfect integration of light and space and try not to create a feeling that light is something extra. There is an interaction between light and mankind, where one can understand and use light.

I would like to take this opportunity to extend my gratitude to Tewa Srilaklang and LDPi!

The best memories are in the heart.

Zhang Jing

鸣谢

经过一个漫长艰巨的过程，成都来福士广场终于落成了。非常感谢LDPi为此项目所作出的努力和贡献。
——成都来福士广场

LDPi为海南香水湾君澜度假酒店进行的灯光设计，完美地将灯具与酒店的自然环境融合到一起，全貌上非常具有整体的设计感。在灯源的设计上，既考虑到了照明的功能性，又体现了灯具本身的美观和装饰感，尊重建筑与环境本身的寓意，用光影凸显了空间的表现力，并在节能、维护和环保问题上都进行了认真考虑。
——海南香水湾君澜度假酒店

我们在北京来福士项目中从LDPi了解并学习到了灯光测试和评价的方法。
——凯德置地

在项目前阶段的合作过程中，双方备感愉快。
——绍兴咸亨新天地酒店

沟通、协作、配合方方面面及时准确。
——冠城名敦道

与LDPi在成都来福士广场项目中的合作可谓极其顺畅。Steven Holl Architects 与 L'Observatoire International 所提出的照明理念需要一个专业素养极高并且敬业的照明设计师团队来实施并保证质量。LDPi的专业性、高度配合以及对行业的了解正是此项目不可或缺的无价资产。
很期待在未来能再次与LDPi合作。
—— Steven Holl Architects

The Buchan Group 是一个全球化的建筑设计事务所，作为商业项目设计的领导者，享有很高的国际知名度，为高水准的国际项目提供设计服务，包括中国的一些地标性建筑。我们可以独自负责项目也可以作为一个比较大的设计团队与其他国际事务所合作提供商业项目设计服务。The Buchan Group 理解灯光设计对客户体验和商业环境的重要性，认为灯光设计应该作为整体商业项目设计的一部分，而不仅仅是一种服务或是附加内容。考虑到照明对建筑的重要性，从项目的早期概念设计阶段，The Buchan Group 就直接邀请照明设计师参与到每个项目的设计中去。
我们寻找最好的有本土经验的照明设计专家合作为客户提供卓越、创新和专业的服务。LDPi 就是一家这样的专业照明设计公司。包括成都来福士在内，The Buchan Group和LDPi已经在很多高水准地标性的商业项目上有过成功的合作。
—— The Buchan Group

Acknowledgements

It's been a long and arduous process for one and all for the coming together of Raffles City Chengdu. A big thank-you is due to LDPi team for the hard work and dedication that is put in.
—— CapitaLand Raffles City Chengdu

The lighting design from LDPi for Narada Resort & Spa Perfume Bay perfectly combines the light with the natural surroundings of the hotel, offering a harmonious integrated whole picture. The design of lighting sources not only takes the functional lighting into consideration, but also showcases the beauty and decorative nature of the devices and respects the architecture and the connotation of the environment. The interaction of light and shadow enhances the expression of spaces. Issues such as energy saving, environmental protection and easy maintainance were also considered by LDPi.
— Narada Resort & Spa Perfume Bay

We have gained valuable knowledge in testing and evaluating lights from LDPi in the CapitaLand Raffles City Beijing project.
—— CapitaLand

We had a very delightful cooperation with LDPi in this project.
—— Shaoxing Xianheng Hotel

Communication, cooperation and coordination in this project are all in punctuality and accuracy.
—— Midtown

Working with LPDi for the Sliced Porosity Block (Raffles City Chengdu) resulted in a very successful collaboration. Key to the success of the lighting concepts by Steven Holl Architects and L'Observatoire International was to count on a strong and dedicated team of lighting designers to ensure the quality of their implementation. LDPI's professionalism, responsiveness and vast knowledge of the trade were indispensable and valuable assets throughout the process.
We look forward to similar collaborations in the future.
—— Steven Holl Architects

The Buchan Group is a global architectural practice and an internationally recognised leader in retail design with a history of engagements to provide architectural services for high profile international projects, including several landmark developments in China. We work in a variety of capacities as either the lead architect or as part of a larger design team working in collaboration with other international practices and providing the retail and mall design. The Buchan Group understands the importance of the lighting design to the customer experience and the retail environment. We regard lighting as an integral part of the retail design, not just as a service or an add-on. Given the importance of lighting to the architecture, The Buchan Group will typically directly engage the lighting designer for each project and we work closely with that designer from the earliest concept stages. We seek out only the very best designers and local experience to collaborate with and choose lighting specialists with a commitment to excellence, innovation and professionalism. LDPi is one such firm. The Buchan Group and LDPi have successfully collaborated on a number of high profile, landmark retail projects in China, including Raffles City Chengdu.
—— The Buchan Group

目 录

序
鸣谢

第一章 酒店

项目	主题	页码
北京香格里拉酒店	境界	16
宁波国际贸易展览中心一期泛太平洋大酒店及公寓	无价	26
昆山阳澄湖费尔蒙度假酒店	夜品美光	32
青岛金沙滩希尔顿酒店	哇！	42
海南香水湾君澜度假酒店	隐藏	48
宁夏银川假日酒店	密码？	56
济南山东大厦酒店	重现	62
大禹·开元精品酒店	夜游	66
绍兴咸亨新天地酒店	献礼中国文学	76
宁波开元名都大酒店	少即是多	86
北京广西大厦酒店	点亮广西文化	92
南昌洗药湖避暑山庄	水雾飞鸟	96
苏州新城花园酒店	飘带	104

第二章 商业综合体

项目	主题	页码
宁波和义大道购物中心	平面、直线和圆弧	114
成都来福士广场	流畅	122
北京金隅万科广场	梦幻宝盒	132
广州太阳新天地购物中心	惊喜！	142
北京来福士中心	"会呼吸"的建筑	150
成都凯丹广场	蓝天之外	158
北京冠城名敦道综合体	从下到上	162

第三章 办公空间

项目	主题	页码
北京凯德置地大厦（现北京SK大厦）	白与金	170
北京宝钢大厦	色温	174
北京第五广场（现中国五矿大厦）	三合一	178
成都中海南城1号B地块1期	翱翔	182
北京亮点设计中心	经验教训	184
宁波市东部新城行政办公楼	保守与现代	194
哈尔滨哈西发展大厦	旋律与音符	200
贵州花果园艺术中心	山中的王国	206
LDPi北京办公室（2007—2012）	"你好，再见！"	212

第四章 住宅空间

项目	主题	页码
北京万科大都会住宅	当灯光遇见奢华	222
北京红玺台住宅区	融合	230
北京瑞士公寓	欧式经典	236
北京华远九都汇	见与不见	240
北京丽都水岸	国画	244

第五章 文化教育及其他

项目	主题	页码
北京万科如园	光与影	248
固安规划馆	2^4	254
上海长颈鹿美语学校	基石	260
成都东站	白加黑	266

Preface
Acknowledgements

Chapter 1 Hotel

Project	Theme	Page
Beijing Shangri-La Hotel	Realm	16
Ningbo International Trade & Exhibition Center Phase I Pan Pacific Hotel and Apartments	Priceless	26
Kunshan Fairmont Yangcheng Lake Grand Hotel	Lighting Delicacy	32
Qingdao Hilton Golden Beach Hotel	Wow!	42
Hainan Narada Resort & Spa Perfume Bay	Hidden	48
Ningxia Yinchuan Holiday Inn Hotel	Code?	56
Jinan Shandong Hotel	Reveal	62
Dayu Kaiyuan Boutique Hotel	Night Journey	66
Shaoxing Xianheng Hotel	Tribute to Chinese Literature	76
Ningbo New Century Grand Hotel	Less Is More	86
Beijing Guangxi Hotel	Switch on Guangxi Culture	92
Nanchang Green Lake Resort	A Bird in the Mist	96
Suzhou New City Garden Hotel	Ribbon	104

Chapter 2 Retail and Mixed Use

Project	Theme	Page
Ningbo Heyi Avenue Shopping Center	Planes, Lines and Arcs	114
Chengdu CapitaLand Raffles City	Fluid	122
Beijing Jinyu Vanke Plaza	Magic Treasure Box	132
Guangzhou Happy Valley	Surprise!	142
Beijing CapitaLand Raffles City	'Breathing' Building	150
Chengdu GTC Galleria	Out of a Blue Clear Sky	158
Beijing Midtown	Bottom-up	162

Chapter 3 Office

Project	Theme	Page
Beijing CapitaLand Tower (Now the Beijing SK Tower)	White and Gold	170
Beijing Baosteel Tower	Color Temperature	174
Beijing Fifth Square (Now the China Minmetal Plaza)	Three-in-One	178
Chengdu China Overseas Property South City Lot1B Phase 1	Soaring	182
Beijing Liang Dian Design Center (LDDC)	Lessons Learnt	184
Ningbo Eastern New Town Administration Zone	Conservatively Contemporary	194
Harbin Haxi Development Building	Melody and Note	200
Guizhou Huaguoyuan Art Center	King of the Hill	206
LDPi Beijing Office (2007-2012)	'Hello, Goodbye!'	212

Chapter 4 Residence

Project	Theme	Page
Beijing Vanke Metropolis	When Light Meets Luxury	222
Beijing Eothen	Fusion	230
Beijing Regent Land	European Classics	236
Beijing Huayuan Jiuduhui	See but Not Seen	240
Beijing Richmond Park	Traditional Chinese Painting	244

Chapter 5 Cultural Space and Others

Project	Theme	Page
Beijing Vanke Ruyuan	Light and Shadow	248
Gu'an Planning Exhibition Center	2^4	254
Shanghai Giraffe English Training Center	Cornerstone	260
Chengdu East Railway Station	White and Black	266

酒店 Hotel

Chapter 第一章

01

北京香格里拉酒店
宁波国际贸易展览中心一期泛太平洋大酒店及公寓
昆山阳澄湖费尔蒙度假酒店
青岛金沙滩希尔顿酒店
海南香水湾君澜度假酒店
宁夏银川假日酒店
济南山东大厦酒店
大禹·开元精品酒店
绍兴咸亨新天地酒店
宁波开元名都大酒店
北京广西大厦酒店
南昌洗药湖避暑山庄
苏州新城花园酒店

Beijing Shangri-La Hotel
Ningbo International Trade & Exhibition Center Phase I Pan Pacific Hotel and Apartments
Kunshan Fairmont Yangcheng Lake Grand Hotel
Qingdao Hilton Golden Beach Hotel
Hainan Narada Resort & Spa Perfume Bay
Ningxia Yinchuan Holiday Inn Hotel
Jinan Shandong Hotel
Dayu Kaiyuan Boutique Hotel
Shaoxing Xianheng Hotel
Ningbo New Century Grand Hotel
Beijing Guangxi Hotel
Nanchang Green Lake Resort
Suzhou New City Garden Hotel

第一章 酒店 | Chapter 1 Hotel
北京香格里拉酒店 | Beijing Shangri-La Hotel

名称：北京香格里拉酒店
地点：中国 北京
完工时间：2004
客户：香格里拉酒店集团
照明设计：英国莱亭迪赛灯光设计合作者事务所 –中国分部

Name: Beijing Shangri-La Hotel
Location: Beijing, China
Year of Completion: 2004
Client: Shangri-La Hotels and Resorts
Lighting Design: Lighting Design Partnership International (LDPi) China Branch

照明设计的故事
Lighting Design Stories

境界

在香格里拉大堂吧的照明设计中可以看到，着重强调的顶部是视觉中心，作为这部分空间主体的吊灯成为关注的焦点；富有文化气息的家具及陈设在统一的光影格调中，演绎出耐人寻味的看点，照明设计师通过"减少笔墨"的手法，使光影成为陪衬，烘托出主题的专一性，所以尽管环境中充满各种元素和色彩，但仍然

Realm

The top of the lobby lounge of the Beijing Shangri-La Hotel is stressed to be the visual center by the application of lighting design. The chandelier, as a principal part, attracts most of the attentions. Standing in the harmonized lights and shadows, furniture and decors rich in culture are

呈现出整体性。

餐厅吊灯周围的环形装饰刻有独特的花纹，在光影的投射下，单调的灯具也散发出绵柔的气息，形成别具一格的装饰效果。灯具在这里既照亮了空间又美化了自己。餐厅是演绎戏剧性的最佳场所：装饰性灯具和功能性灯具相互配合，使环境主题得以彰显；贵重奢华的餐桌经过处理，呈现恰到好处的纹理和色彩，且不会产生

expressive and meaningful. Here, lights as a part of the background help to create a single-themed environment. Therefore, despite being filled with different elements and colors, the lounge is still perceived as an integrated space.

In the restaurant, around the chandelier, one can notice a ring-shaped

Lighting Design Stories

明显的反射眩光,在光影的照耀下与即将登场的美食形成一方舞台,而围坐的宾客就像是观众;暗色的座椅吸收了大部分光线,只等尊贵的客人落座才显现出蓬荜生辉的光彩。

在客房中,环境中的元素通过材质的"硬"体现家纺的"软",通过周边光色的"冷"体现居室中心的"暖",通过"有形"之光(灯具光)与"无形"之光(自然光)

decoration carved with unique patterns. With the projection of lights, this simple lighting facility suddenly becomes gentle and unusually decorative. The chandelier is then not only an illuminant but also an ornament.

No other places than restaurants are best to show the essence of theatricality. Lights, either decorative or functional in nature, are well decorated

第一章 酒店
北京香格里拉酒店 | Chapter 1 Hotel — Beijing Shangri-La Hotel

的对比，确立床榻的中心地位。在这里你看不到多余的光线，这种抛弃累赘的手法使环境氛围更加安逸。

值得一提的是，越是好的照明设计，光的表现就越是柔和、隐蔽。一个优秀的环境设计是让体验者全身心地投入体验人与环境的关系，而不是被环境包围、俯视和压迫。光的参与在功能上像一个老练沉稳的侍者，在你需要的时候现身为你服

to highlight the mood of the space. With smart lighting design, those luxurious dining tables present the right patterns and colors, no more and no less, without any obvious reflected glares. It seems, under the lights, tables and kickshaws are on stage, and guests are the spectators. Dark-colored chairs absorb most of the lights, yet when people are seated, the

务，不需要的时候则隐身遁形。在艺术格调上，好的照明增加了艺术的价值与氛围，但照明本身仍然退居其后，高明的照明设计甚至会引导你的目光超越艺术品本身，从而关注艺术价值的所在。

值得玩味的是：高明的照明设计使你忘记设计师的工作，而这样的设计才堪称大师之作。应该说照明设计的不是光，而是一种境界。

chairs then become lively and brightened up.

In the guest rooms, the position of the bed is accentuated through several contrasts. Formless natural light and the form of artificial lighting create warmness background of room. With no redundant light, this design makes the whole room more comfortable as the hardness of the elements

第一章 酒店
北京香格里拉酒店 | Chapter 1　Hotel
Beijing Shangri-La Hotel

in the general surrounding emphasizes the softness of the textiles.
The better a lighting design is, the more tender and subtler the light is. The core of an outstanding environmental design is to let visitors dedicate themselves to experiencing the relationship with the environment, rather than being surrounded and pressed by the environment. Lighting in functional perspective, just like an experienced servant, is always there when needed and disappears when unnecessary. In artistic perspective, good lighting increases the value and ambience subtly without any attracted attention on lighting itself. A smart lighting designer may even guide people to go beyond works of art to learn the real value of art.
In a clever design, audiences often enjoy the lighting without realizing what they are witnessing is the work of designers. And this is the true masterpiece. Lighting design is a state of art rather than just about lights.

名称：宁波国际贸易展览中心一期泛太平洋大酒店及公寓
地点：中国宁波
完工时间：2012 年
总建筑面积：170 000 平方米
客户：宁波市国际贸易投资发展有限公司
建筑设计：Spark（原 ALSOP/SPARCH）
照明设计：英国莱亮迪赛灯光设计合作者事务所－中国分部

Name: Ningbo International Trade & Exhibition Center Phase I Pan Pacific Hotel and Apartments
Location: Ningbo, China
Year of Completion: 2012
Gross Floor Area (GFA): 170,000 m²
Client: Ningbo International Trade Investment Development Co., Ltd.
Architectural Design: Spark (formerly ALSOP / SPARCH)
Lighting Design: Lighting Design Partnership International (LDPi) China Branch

第一章 酒店 | Chapter 1 Hotel
宁波国际贸易展览中心一期泛太平洋大酒店及公寓 | Ningbo International Trade & Exhibition Center Phase I Pan Pacific Hotel and Apartments

无价

宁波国际贸易展览中心一期项目包括泛太平洋大酒店、服务式公寓以及五栋底层带有零售商铺的公寓住宅，坐落在东部新城的开发区内，世纪大道与民安东路东南交汇处，周围是国际金融中心、宁波国际会展中心等地标性建筑。距离市中心仅10分钟车程，距离栎社机场40分钟车程，交通异常便利。

Spark设计的建筑外墙融合了现代元素，充分契合了宁波东部新城区成为新政治、经济、文化和商业中心的目标。泛太平洋大酒店建筑使用不同处理手法呈现出玻璃与石材材质的混搭效果，结合黑白横向"条形码"的建筑理念，使整栋酒店呈

Priceless

Ningbo International Trade & Exhibition Center Phase I project includes a Pan Pacific Hotel, a serviced suites and 5 residential buildings with retails on the ground floor. The project lies in the development zone of Ningbo Eastern New Town (ENT) and the Administrative Zone, neighboring International Financial Center and Ningbo International Exhibition Center, at the southeast intersection between the Century Avenue and East Min'an Road. It is a mere 10-minute drive from the city center and

现简洁而又有韵律感的风格特色。照明设计师敏感地抓住了这种特点,并在照明设计上加强体现了这一特点:夜幕降临,暗暗的石材与打亮的玻璃相间,它们在甬新河上的倒影吸引着众人的目光。设计师在玻璃石材的窗间层布设了4 000 K的LED灯带,同时使房间有效地避免光扰。为了不影响客人休息,布光多设置在临街临河的外立面,内侧则减少用光。整体建筑照明设计采用了可调光系统,增加了酒店的活力与情趣。

位于酒店北面的五座公寓楼,其建筑结构和材质与酒店相同,也同样采用了"条形码"这一理念,只在外围建筑立面窗层的拐角处用3 000 K暖白光打亮,最大

40-minute drive from the Lishe airport.

Being in Ningbo's ENT, Spark's architectural design of the façade includes contemporary elements matching the goal of being the new political, economic, cultural and commercial center. The architectural design combines glass and stone forming a simple and rhythmic black and white horizontal 'barcode' during the day. The lighting design accurately catches such unique barcode feature and takes it into another dimension by making it doubly complementary at night: while the stone remains in the

第一章 酒店
宁波国际贸易展览中心一期泛太平洋大酒店及公寓

Chapter 1 Hotel
Ningbo International Trade & Exhibition Center Phase I Pan Pacific Hotel and Apartments

照明初步概念图
Preliminary lighting concept rendering

限度地保证隐私和静谧程度。入夜，公寓楼精心设计的照明，简洁大气，给人以愉悦、安静、归家的感觉。

隐藏光源，简单布灯，场景设置以及技术整合，使灯具在白天丝毫不会破坏建筑形式和设计理念，这正是设计团队在每一个照明设计项目上都力争达到的效果。夜幕降临，隐藏起来的充满戏剧性的灯光延续了建筑的生命，灵动的光效好似正在被扫描器阅读的条形码，显示出它的无价设计。

dark, the glass glows and furthermore, both reflect on the Yongxin River and appeal to the public's attention. To achieve this effect, the team chose to have a clean white 4,000 K LED linear light installed along the glass windows accessible for maintenance from above the guestroom's curtains, thus avoiding any light disturbance into the room. With the same consideration, the team chose to keep a comfortable lighting environment for guests and designed the inner façade with less lights installed whereas the facades facing streets and the river are equipped with more lights.

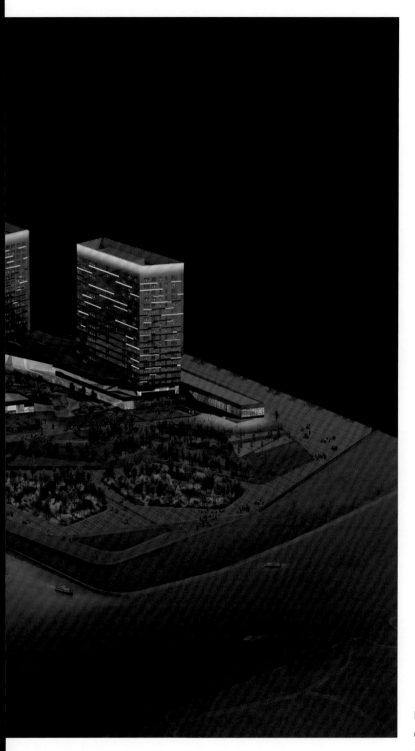

In addition, the team required the lighting fixtures to be dimmable via a central controlling system, adding vividness and dynamism of the overall effect.

The five apartment buildings are located north of the hotel and are designed with the same building materials and structure. There, still keeping the 'bar code' concept in mind, the team chose to change the color temperature, having a 3,000 K warm white lights installed only at the outer corners of the facades of the outer buildings to preserve the privacy and quietness of the residents. At night, the well arranged lighting of the apartment buildings follows the simplicity of the architectural design, quietly creating a joyful home-coming spirit.

Achieving and complementing the architectural design without compromising the daytime look makes the lighting design a subtle consulting service that the team strives for in all projects, including this one. Smartly using a simplistic design with modern technology transforms this sparkling bar code read by scanners into a priceless design.

第一章 酒店 | Chapter 1 Hotel
昆山阳澄湖费尔蒙度假酒店 | Kunshan Fairmont Yangcheng Lake Grand Hotel

名称：昆山阳澄湖费尔蒙度假酒店
地点：中国昆山
完工时间：2008 年
总建筑面积：10 566 平方米
客户：昆山城市建设投资发展有限公司
室内设计：BENT SEVERIN & ASSOCIATES PTE LTD
建筑设计：苏州市规划设计研究院有限公司第一分公司
照明设计：英国莱亭迪赛灯光设计合作者事务所－中国分部

Name: Kunshan Fairmont Yangcheng Lake Grand Hotel
Location: Kunshan, China
Year of Completion: 2008
Gross Floor Area (GFA): 10,566 m²
Client: Kunshan City Construction Investment & Development Co., Ltd.
Interior Design: BENT SEVERIN & ASSOCIATES PTE LTD
Architectural Design: Suzhou City Planning and Design Institute Ltd. (First Branch)
Lighting Design: Lighting Design Partnership International (LDPi) China Branch

夜品美光

提起苏州东北部美玉般嵌于昆山之中的阳澄湖,自然会让人想起秋日里鲜香甜美的阳澄湖大闸蟹。有美食闻名,自然会吸引远方的客人前来一探究竟,品其味、观其景,让这里的酒店餐饮业日益兴盛。阳澄湖畔的费尔蒙酒店则是目前周边现有的唯一一家能坐览如此湖畔美景的五星级酒店,从建筑结构、室内设计及灯光设计的角度,老道地刻画出这江南式美景,盛名贵于精巧,值得细品一翻。

酒店建筑高7层(不包括地下2层),整体结构呈现一个巨大的弧形。酒店公共区域主要分布在1层及地下1层,2层以上均为客房,共计200间。这个特点使

Lighting Delicacy

Like a fine jade embedded in Kunshan in the northeast of Suzhou, Yangcheng Lake is famous for the local delicacy, hairy crabs. Each year tourists come from afar for the food and the beautiful scenery, making the local hospitality industry flourishing. By the side of Yangcheng Lake, the Fairmont Yangcheng Lake Grand Hotel is currently the only 5-star hotel that has the privilege to embrace the fine lake view. Its building structure, interior design and lighting design smartly nail the southern beauty and

第一章 酒店 | Chapter 1 Hotel
昆山阳澄湖费尔蒙度假酒店 | Kunshan Fairmont Yangcheng Lake Grand Hotel

酒店的每间客房都拥有绝佳的全景观赏视角。从酒店客房内便可一眼望见窗外湖畔散布的点点粉墙黛瓦人家，湖面依稀可见的农家蟹网，带着江南烟雨的柔婉，正应了江南水乡的气质。对这样一个江南特色酒店，照明设计师希望通过灯光设计，将室外江南美景延伸至室内空间，并为客人营造出宁静、舒适的体验环境。

费尔蒙酒店室内空间设计的细节处理非常精彩，尽量保持开阔的整体感及流线型的设计安排，同时在设计中巧妙地融入了湖边的水草、芦苇等带有地缘特征的景物。这些环境因素与吹制玻璃结合制成灯具装置，或构成装置水景，不露声色地呈现了地域美景，让人耳目一新。基于室内设计师对空间的把握，照明设计师则

are worth a careful study.

The 7-storey arc shaped hotel (excluding 2 floors underground) is designed with its public areas of the hotel on the ground floor and a floor below. The 200 hotel rooms start from the second floor and offer guests a unique lake view. Looking out from the windows, one can see folk housing with white walls and black tiles scatter by the lakeside. On the lake, crab nets and gentle misty rain jointly interpret the essence of canal towns in southern China. The lighting design of Fairmont Yangcheng Lake Grand

通过灯光设计在夜晚还原出白天的氛围，并体察室内设计师背后的思考，以主动创造的姿态参与其中，利用光的语言传达出室内和室外环境的联系。

照明和以自然为主题的内饰在酒店的许多空间中巧妙融合。大堂的照明被分为三层，引导宾客前行。前台背倚绿色环状玻璃幕墙首先映入眼帘，清新淡雅的背景灯打在上面营造出悦目的氛围。水晶玻璃天花板和水景池采用较弱强度照明，在大堂中心交相呼应。这三个圆形元素好似环形波浪，使宾客仿佛置身于阳澄湖之中。

费尔蒙酒店SPA空间中，水已然是一个很重要的组成部分，照明设计则将重点转

Hotel aims to bring the beauty of the local scenery to the interior of the hotel, creating a quiet and comfortable ambiance.

The interior design of this hotel pays careful attentions to details. Not only the overall integrity and streamlining design are maintained, but also objects with the features of local aquatics and reeds are integrated into the overall planning. Those environmental factors and blown glasses are turned into special lighting devices and artificial waterscapes, quietly showcasing the local beauty and bringing a refreshing look. Based on

VIP 楼
VIP building

第一章 酒店
昆山阳澄湖费尔蒙度假酒店
Chapter 1 Hotel
Kunshan Fairmont Yangcheng Lake Grand Hotel

向其他元素，将人造的室内环境打造成室外的自然美景。SPA 的室内设计以"竹"为主，照明设计则用光来凸显这一主题。设计师将光导纤维散布在天花板上，仿佛繁星密布的天空，营造出一种惬意的自然环境。

费尔蒙酒店的 Curve 酒吧格外引人关注，它有着适合轻松交谈的氛围，雅致却精心，是私密约会或品酒社交的绝佳场所。这个空间的天花水晶吊灯很是引人瞩目，辅以错落有致的光点排布，向上仰望时，好似大珠小珠落玉盘。晶莹、闪烁、梦幻的光影，对应脚下水流花纹的地毯，奢华而不张扬。

费尔蒙酒店另一功能区——金酒廊的照明设计与酒廊有异曲同工之妙。由波浪形

the interior design, the lighting designer creates a day view at night in the hotel and connects the interior and exterior via active lighting.

In the several location in the hotel, lighting blends with other natural themed interior elements. When entering the hotel, guests are guided through the lobby with three layers of lighting. The first visible element is the front desk, which seats in front of a ring shaped greenish glass wall, refreshingly back lit. The next two elements, with a lower lighting intensity are the decorative ceiling glass blown light and the waterscape in the

吊顶中打下的灯光使地毯上波纹形状的花纹显得格外突出。
阳澄湖费尔蒙酒店的独特地理位置和景色使其成为独一无二的五星级酒店。除了美味的大闸蟹外,那些以水为主题的照明设计组成了一桌饕餮的视觉大餐。分布在费尔蒙酒店走廊及宴会厅等地方的水草造型的装饰灯,和点缀其中的昆山文化艺术品,都在告诉客人这里是阳澄湖,是美丽运河城市的典范。

middle of the lobby. All elements are round shaped in a way of wave ring water droplets as if guests were in the midst of the Yangcheng lake. As water is already present in the Spa in this hotel, lighting has to focus on other elements to transform an artificial interior into a natural exterior environment as the interior designer envisioned bamboo being one of the key theme elements, lighting could only make it stand out even more. To complement this natural element, optical fiber is added to the ceiling above the SPA so that guests feel relaxed under a starry sky.

Chapter 1 Hotel
Kunshan Fairmont Yangcheng Lake Grand Hotel

There is no better place than the Curve bar for private and business socializing. This place is filled with comfort and elegance. The crystal ceiling light neatly easily attracts attentions, as if it were raining pearls from high above the sky onto the carpet with water-wave patterns bring out a feeling of natural luxury.

In a similar way, another functional area of the hotel is the Fairmont Gold Lounge, where guests see another type of wave centered pattern on the carpet made visible by beams of lights hidden in between wavy ceiling

screens.

With a unique location and views, the Fairmont Yangcheng Lake Grand Hotel is no ordinary a five-star hotel. Delicacies are found not only in the form of the hairy crabs throughout the lake, but also in the water related interior design theme cascading down into a pond of lighting delicacies.
The aquatic-shaped lights in corridors and banquet halls and local artifacts not only jointly remind guests they are at Yangcheng Lake but also showcase the beauty of canal towns.

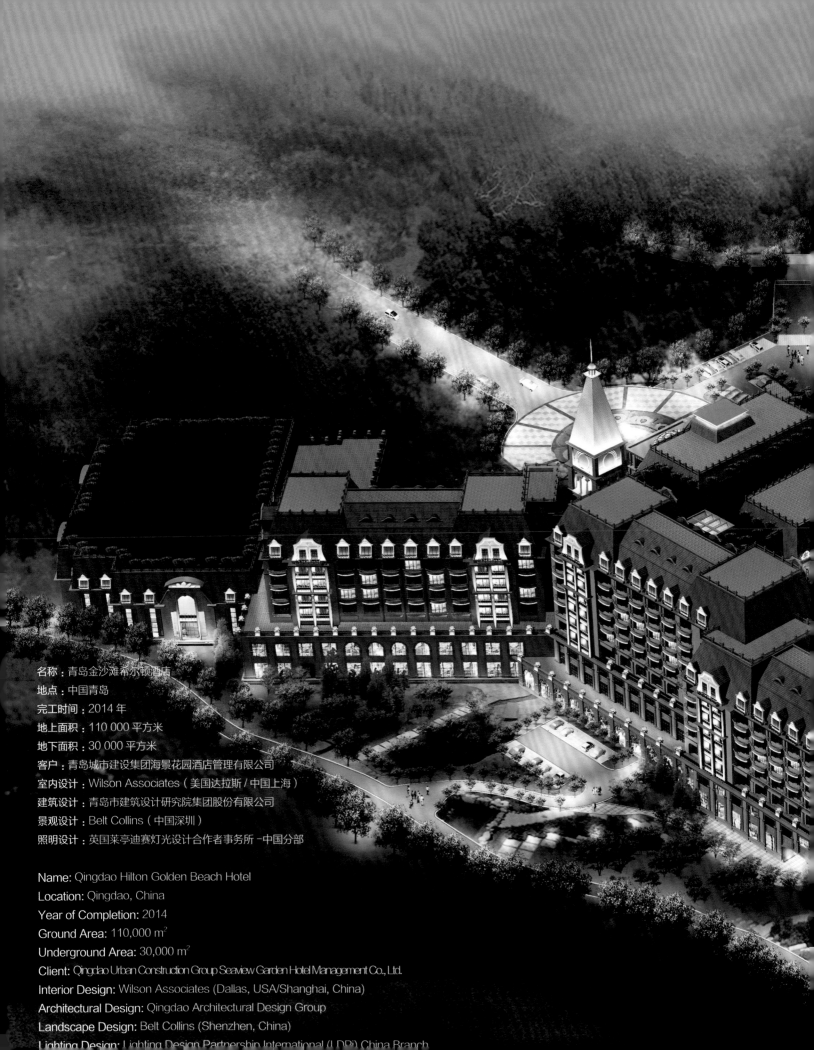

名称：青岛金沙滩希尔顿酒店
地点：中国青岛
完工时间：2014年
地上面积：110 000 平方米
地下面积：30 000 平方米
客户：青岛城市建设集团海景花园酒店管理有限公司
室内设计：Wilson Associates（美国达拉斯 / 中国上海）
建筑设计：青岛市建筑设计研究院集团股份有限公司
景观设计：Belt Collins（中国深圳）
照明设计：英国莱亭迪赛灯光设计合作者事务所 -中国分部

Name: Qingdao Hilton Golden Beach Hotel
Location: Qingdao, China
Year of Completion: 2014
Ground Area: 110,000 m²
Underground Area: 30,000 m²
Client: Qingdao Urban Construction Group Seaview Garden Hotel Management Co., Ltd.
Interior Design: Wilson Associates (Dallas, USA/Shanghai, China)
Architectural Design: Qingdao Architectural Design Group
Landscape Design: Belt Collins (Shenzhen, China)
Lighting Design: Lighting Design Partnership International (LDPi) China Branch

照明初步概念图
Preliminary lighting concept rendering

第一章 酒店 | Chapter 1 Hotel
青岛金沙滩希尔顿酒店 | Qingdao Hilton Golden Beach Hotel

照明初步概念图
Preliminary lighting concept rendering

哇！

青岛金沙滩希尔顿酒店位于青岛经济技术开发区，与黄岛区金沙滩景区一路之隔，距青岛市中心20公里，驱车1小时便可抵达机场。金沙滩希尔顿酒店欧式风格的整体建筑结合了现代室内设计和奢华的景观设计，毫无疑问是一个建筑杰作。在2010春末，酒店照明设计伊始遇到了两处巨大的限制。首先，建筑的地基已经完成照明顾问才介入项目，对幕墙进行灯光设备布位设计及细节安装，并安排合理的走线方案；其次，由于天气原因，冬天通常会停工，考虑到这两点，设计团队合理规划时间赶在冬天来临之前备齐所有关于幕墙的设计内容，还同时进行

Wow!

Qingdao Hilton Golden Beach Hotel is just a road away from the soft, wide sands of Golden Beach on the Yellow island, located in Qingdao's Economic and Technology Development Zone (ETDZ), 20 km from downtown Qingdao and about an hour away from the airport. Qingdao Hilton Golden Beach Hotel is an architectural masterpiece. It combines an European style architecture, a modern interior design and luxurious landscape design.

Lighting Design Stories

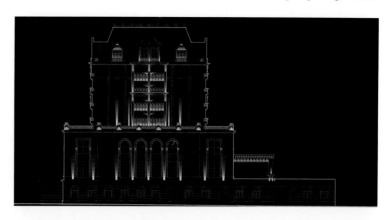

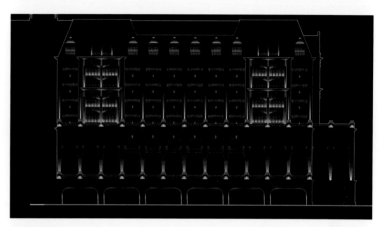

照明初步概念图
Preliminary lighting concept rendering

景观方案的深化设计。

在项目初期与Wilson Associates达拉斯工作室的负责人James Carry会面时，LDPi做了一次别出心裁的方案设计汇报，不仅囊括了每个空间的功能，还涉及了空间将来的运行和操作，将酒店对于照明的期待和要求表达得清楚明白，比如，厨房应当离餐桌近一些，这样才能保证用餐时菜品的温度恰到好处。

当踏入酒店大堂，宾客们会被众多惊艳的元素吸引住，不由自主地发出赞叹。气派的大堂映入眼帘好似一场绚丽的梦境。在最初的设计中，客人们不仅会对宽大的酒店空间、舒适的环境和便利的设施赞不绝口，还会被大堂中心四层楼高的斗

When the team started the lighting design for this project before the summer of 2010, there were two huge constraints. First, the foundations of the building were already completed and the curtain wall was waiting for the lighting consultant to join the project in order to provide the lighting fixture location and detailed installation to prepare ducting for the electrical wiring. Secondly, due to harsh weather conditions, the construction is usually put on hold during winter time. With those two facts in mind, the project team allocated time to catch up with the timeline to provide all nec-

第一章 酒店
青岛金沙滩希尔顿酒店 | Chapter 1 Hotel
Qingdao Hilton Golden Beach Hotel

照明初步概念图
Preliminary lighting concept rendering

拱形设计及其两边矗立的四个两层楼高的手工玻璃艺术品震撼，这些艺术品向下弯曲呈鞠躬状，并有着特殊的照明设计。这种特殊设计对于玻璃艺术品的质量有着很高的要求，必须要达到欧洲标准，并且灯光设计也必须能够精确地抓住人们的眼球。然而，由于资金的限制，当地的玻璃制造商不能提供出高质量的作品，因此他们提供了其他的可装饰酒店大堂的精美作品供 Wilson Associates 选择，毋庸置疑，这将会成为酒店的另一点睛之笔，同样，照明设计也不例外。

essary information for the curtain wall to proceed before the winter kicked in, from the schematic design to the design development. At the same time it worked on the landscape schematic design and design development.

When the interior design started, during the first project meeting with James Carry, the principal of Wilson Associates' Dallas office, LDPi made an interesting schematic design report. Elements not only included the functions of each space, but also the operations of the space. For

照明初步概念图
Preliminary lighting concept rendering

instance, the kitchen could not be too far from the table where the food would be delivered in order to preserve optimal serving temperature. Expectations and requirements related to lighting were clearly expressed and understood.

When customers entered the lobby, the 'Wow!' factors captivated their attention. Just like a good salesman, a dream was sold right on the spot. In the original design, the dream and unique feature of this hotel would be that not only guests would be impressed by the size of the hotel and its amenities, but also guests would be welcomed into the hotel lobby by four two-storey high bowing hand-blown glass art work pieces that would require specific lighting treatments under the four-storey high arched shaped central part of the lobby. The expectations for these four art works were set high both for the glass quality (of European standards) and for the lighting that must bring a striking view to the guests.

As of the writing of this paper, due to budget constraints, a local glass manufacturer for the glass art work could not deliver the expected quality, bringing other considerations for the customer and Wilson Associates to choose as the central master piece of the hotel. Without any doubt, it will be another 'Wow!' factor and so must be the lighting.

第一章 酒店
海南香水湾君澜度假酒店 | Chapter 1 Hotel — Hainan Narada Resort & Spa Perfume Bay

名称：海南香水湾君澜度假酒店
地点：中国海南
完工时间：2011 年
占地面积：110 000 平方米
总建筑面积：33 680 平方米
客户：海南香水湾君澜度假酒店
室内设计：PAL
建筑设计：AAI
景观设计：EDAW
照明设计：英国莱亭迪赛灯光设计合作者事务所 - 中国分部

Name: Hainan Narada Resort & Spa Perfume Bay
Location: Hainan Province, China
Year of Completion: 2011
Site Area: 110,000 m²
Gross Floor Area (GFA): 33,680 m²
Client: Narada Resort & Spa Perfume Bay
Interior Design: PAL
Architectural Design: AAI
Landscape Design: EDAW
Lighting Design: Lighting Design Partnership International (LDPi) China Branch

隐藏

海南香水湾君澜度假酒店位于三亚市东北部，距离繁华的中心地带有 80 公里之遥，这个面朝大海的热带度假酒店拥有酒店客房和私人别墅。

不同于以往，本项目的设计团队从一个独特的视角出发，在项目的概念设计阶段，通过观察和体验，按照宾客办理入住手续一直到自己的房间这一过程来开始最初的设计，而到了设计的深化阶段，还要考虑避免多余的外立面和景观照明遮盖了美丽的自然海滨景色。

室内设计深受酒店外观的启发，延用了东南亚热带自然景观元素。酒店前台后方，

Hidden

Hainan Narada Resort & Spa Perfume Bay is a superb sea front tropical resort 80 km northeast of the central busy and touristic city of Sanya, Hainan. There guests have the luxury to either be hosted in one of the hotel guestrooms or in a private villa.

When designing the lighting for the Narada Resort & Spa Perfume Bay, the team took the chance to make a design from a different perspective. Not only was the schematic design envisioned from a hidden view

第一章 酒店
海南香水湾君澜度假酒店 | Chapter 1 Hotel
Hainan Narada Resort & Spa Perfume Bay

矗立着一个两层楼高的洁白的羽毛墙，墙上浮雕三枝天堂鸟花，时时刻刻欢迎着远道而来的宾客。暖白的LED则隐藏在浮雕下。虽然在白天强烈的日光下，天堂鸟花的灯光依稀可见，然而只有到了夜晚，它们的美连同整体墙面的设计方才淋漓尽致地展现出来。

有着天堂鸟花造型灯罩的立式灯引导着客人步入房间，从而使这一元素贯穿公共空间和私密空间，为宾客带来愉悦之感。

放下行囊，客人们可以全身心地享受美景与酒店提供的服务。中轴路的尽头是无尽的大海，路两旁种满椰树，夜晚时分灯光将椰树打亮。酒店大堂延伸出来的水

amongst a group of guests going through the resort after the hotel check-in process, but also the design development ensured that the superb views weren't blocked by external lighting or lighting from the landscape. The interior design was strongly inspired by the luxuriant tropical landscape elements of this Southeast Asian resort. In the hotel lobby, behind the check-in counter, sits a two-storey high clean white feature wall. There, three bird-of-paradise flowers are carved out from the wall, welcoming guests at any time of the day for their check-in and check-out

照明设计的故事
Lighting Design Stories

景穿过酒吧，汇入海滩上的游泳池。在夜晚舒适的灯光照耀下，客人们尽情地享用着美食和美酒，不受外界灯光的打扰。
酒店屋檐的照明设计费尽心机：照明装置被巧妙地藏匿起来，形成非常有序的水平光束，将房顶的独特设计和材质展现得淋漓尽致。
海南香水湾君澜度假酒店远离喧嚣，只为那些寻找独特静谧环境入住的宾客们独享。其照明设计理念也完全依循着隐藏的原则。

procedures. Warm white LED lights were installed and hidden from the public's eyes behind the carving. Although during daytime with strong daylight, the lighting of these three bird-of-paradise flowers is noticeable, it only shows its full beauty at night together with the texture of the background wall.
Decorative free standing lights with bird-of-paradise lamp shade also welcome guests into their rooms, making the flower being enjoyable not only in public areas, but also in private spaces.

第一章 酒店 | Chapter 1 Hotel
海南香水湾君澜度假酒店 | Hainan Narada Resort & Spa Perfume Bay

Once baggage is left in the rooms, guests can fully enjoy the resort. With a full sea view from the central axis with coconut trees lit up by night, and waterscape starting from the back of the hotel lobby, going through the bar area, before transforming itself into a swimming pool lit up at night and stopping on the top of the beach, guests can enjoy beverages and food without being disturbed by any overwhelming exterior lighting.

Indeed, the roof lighting received a special treatment: again the design

was to install the light source hidden and recessed into the structural horizontal beam. Once turned on, the light reveals the unique textures of the roofs.

The Narada Resort & Spa Perfume Bay is a splendid resort hidden away from the crowd that only the ones who are seeking a unique and calm tropical location to stay can enjoy. The lighting there is fully integrated into the resort in line with this hidden principle.

第一章 酒店 | Chapter 1 Hotel
宁夏银川假日酒店 | Ningxia Yinchuan Holiday Inn Hotel

名称：宁夏银川假日酒店
地点：中国银川
完工时间：2011年
占地面积：35 514 平方米
总建筑面积：50 162 平方米
客户：宁夏京能创业房地产开发有限公司
室内设计：北京东方国兴建筑设计有限公司
建筑设计：北京东方国兴建筑设计有限公司
照明设计：英国莱亭迪赛灯光设计合作者事务所-中国分部

Name: Ningxia Yinchuan Holiday Inn Hotel
Location: Yinchuan, China
Year of Completion: 2011
Site Area: 35,514 m²
Gross Floor Area: 50,162 m²
Client: Ningxia Jingneng Chuangye Real Estate Development Co., Ltd.
Interior Design: Beijing Dongfang Guoxing Architectural Design Co., Ltd
Architectural Design: Beijing Dongfang Guoxing Architectural Design Co., Ltd
Lighting Design: Lighting Design Partnership International (LDPi) China Branch

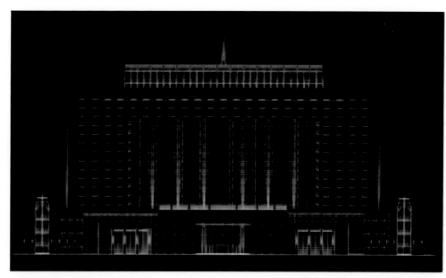

照明初步概念图
Preliminary lighting concept rendering

密码？

宁夏银川假日酒店位于银川市中心中山公园南部。

酒店的照明项目设计之初遇到了一个时间方面的小问题：室内照明设计的时间非常充裕，然而早已竣工的外立面因为没有预先做照明规划，所以在这方面设计的需求却很紧急。在外墙上不能附加任何线路，所以在墙面上安装灯具几乎不可能。因此，设计师通过在酒店周围的景观布光以及在公众触及不到的墙面安装灯具的方式来设计外墙照明。

酒店主楼外墙还有一个非常显著的特点，整体建筑以酒店入口处中心轴线为中心

Code?

The Ningxia Yinchuan Holiday Inn is located right at the center of the city, south of the Zhongshan Park.

When the lighting project was started, there was just a little hiccup: time allocated for the interior lighting was sufficient, whereas the façade was in an urgent need for lighting. Indeed as the façade was already completed but façade lighting was an element which was completely missing. And because no additional wiring was possible from the façade, this left little

第一章 酒店 | Chapter 1 Hotel
宁夏银川假日酒店 | Ningxia Yinchuan Holiday Inn Hotel

对称,这个灵感来源于达芬奇对于罗马建筑师维特鲁维的致敬之作,被誉为对称神作的《维特鲁维人》。

从街道的角度来看,马路东面还有两座较为低矮的酒店,所以设计团队决定重点突出酒店东西两侧。用12个投光灯分别洗亮两侧墙体,线型LED灯带被安装在楼梯间。南立面采用窄角投光灯打亮以减弱照明对客房的干扰。

从人行道角度来看,被完全隐藏的64个宽角投光灯,在主入口上方打造了一个令人惊艳的皇冠。其中6个投光灯分布在两侧,32个壁灯则勾勒出裙楼的整体轮廓。

room of maneuver but to install lighting either on the landscape nearby the building, or attached to the facade in an area not accessible by the public. With that said, the building's main façade has another special and spatial particularity: it is a vertical axial symmetry with an axis located at the center of the hotel entrance. With such feature, the source of the lighting inspiration was found in the tribute paid by Leonardo Da Vinci to the Roman architect Vitruvius with the *Vitruvian Man*, often used as a symbol of symmetry.

照明设计的故事
Lighting Design Stories

尽管你在室内设计中找不到破解达芬奇密码的关键,但是 LDPi 拥有足够多的创意,在与室内设计完美融合的同时,达到酒店管理方对室内设计的标准和要求。

From a street scale, because there are two other lower scale hotels located on the east side, the team chose to highlight the west and east sides with a set of 12 projectors to wash each side and linear LEDs at the staircases; the south elevation is simply lit with projectors: narrow beams on façade to minimize guestroom disturbance.

At the pedestrian level, there are 64 wide beam hidden projectors to make the impressive crown stand out above the main entrance, around which 6 projectors highlight on each side and finally there are 32 wall lamps mark-

第一章 酒店 | Chapter 1 Hotel
宁夏银川假日酒店 | Ningxia Yinchuan Holiday Inn Hotel

ing the shape of the podium.

Although no Da Vinci code related element was found for the interior, LDPi had enough content to absorb in order to design and integrate the interior lighting based both on the interior designer and the hotel management's design code, standards and requirements.

第一章 酒店 | Chapter 1 Hotel
济南山东大厦酒店 | Jinan Shandong Hotel

改造后夜晚景色
Night View after Renovation

名称：济南山东大厦酒店
地点：中国济南
完工时间：2009 年
占地面积：95 000 平方米
总建筑面积：130 000 平方米
客户：济南山东大厦酒店
照明设计：英国莱亭迪赛灯光设计合作者事务所 –中国分部

Name: Jinan Shandong Hotel
Location: Jinan, China
Year of Completion: 2009
Site Area: 95,000 m²
Gross Floor Area: 130,000 m²
Client: Jinan Shandong Hotel
Lighting Design: Lighting Design Partnership International (LDPi) China Branch

重现

2002年3月18日，山东大厦正式落成。酒店位于济南的文化中心历下区，东倚千佛山，北邻泉城公园。没有都市建筑群落中的鳞次栉比与熙熙攘攘，山东大厦的挺拔尤为突出。作为济南市最重要的会议场所之一，山东大厦拥有五星级酒店的资质，曾有幸接待过中国前国家主席胡锦涛。

十几年前，约翰·波特曼建筑设计事务所承接建造山东大厦时，灯光设计的概念在国内并未得到重视。所以，山东大厦的外部灯光系统的安装也受制于当时的技术。随着时间的推移，经过不断的维修保养，大厦外立面的灯光变得杂乱，如无章拼凑的补丁一般，难以展现建筑的气势，也不能满足设计师和业主的需要。

2009年，LDPi承接了山东大厦，这个具有挑战性的改造项目。在最初的实地考察中，项目团队发现了亟待解决的问题：大厦外立面施以大功率、色温各异的泛光灯，未能恰到好处地突出建筑自身特点。考虑到山东大厦的建筑结构特征、地理位置和主要功能，设计师采用现代照明技术，针对建筑特点布光，与此同时避免照亮其他部位，如此将照明与建筑巧妙结合。

山东大厦外立面原有的泛光照明系统不仅未能突出建筑本身的弧形特点以及建筑气势，也使得弧形主楼在夜间变得索然无味。经过对外立面的仔细研究，设计师

Reveal

Shandong Hotel, officially opened on March 18, 2002, is situated in Lixia District right at the cultural center of Jinan, Shandong Province. With Qianfo Hill to the east and Quancheng Park on the north and without crowded surroundings, Shandong Hotel stands tall and strong. As such and being one of the key conference venues in Jinan, Shandong Hotel, achieving 5-star standards, had the privilege to welcome former president HU Jintao.

More than a decade ago, when John Portman & Associates was appointed as the architect of the Shandong Hotel, lighting design did not attract sufficient attentions in China. The exterior lighting of the hotel was installed with the technology that was available at that time. Through the course of time and maintenances, the façade became a mix and match of unflattering lighting patches to the building, its designer and its owner.

In 2009, LDPi took on the challenge to revamp the Shandong Hotel with an appropriate lighting makeup makeover. During the initial site visits,

第一章 酒店 | Chapter 1 Hotel
济南山东大厦酒店 | Jinan Shandong Hotel

改造前夜晚照片
Before renovation

照明初步概念图
Preliminary lighting concept rendering

发现可以在阳台上安装亮度可调的暖白线型 LED 灯具以突出主体建筑轮廓。随着项目的推进，主楼的夜间照明逐渐完善，层次鲜明的照明勾勒出了山东大厦原始美景。

业主的配合在这个改造项目中起到了重要的作用，在重点部位布光，力求凸显建筑特色。为完善会议中心的照明，设计团队在入口顶端安装了横向的暗槽灯，并采用喷砂处理光滑材质巧妙地隐藏了新装灯具。虽然，山东大厦的标志不能施以背光，但却可以运用会议中心两侧的射灯将其打亮。

照明设计不仅是设计灯光，同时也是设计影子。设计师除了对实体灯具的布置、光源、色温和角度的把握，也通过对影子的处理来塑造物体形象、营造空间氛围，达到立体的光环境。在山东大厦的一侧，设计师调整了几株树木植物的光照角度，使其剪影恰到好处地投射在建筑的外立面上，用这种方式将轻柔、婆娑的自然姿态完美地融入了硬朗规则充满现代工业感的几何形态。

在这个改造项目中，设计师用自己对灯光的理解重新解读了山东大厦，并用创造性的手法重新刻画了它的外形。夜幕降临，山东大厦在灯光的映衬下，从鳞次栉比的建筑中脱颖而出，别致而富有生命力。

LDPi team noticed various open points to resolve: inefficient high power flood light projectors washed the façade with different color temperatures, concealing the features of the building. Considering the structure, location and function, the hotel's lighting design mainly focused on finding key architectural features and elements onto which current lighting technology can be applied while keeping the rest unlit, thus integrating lighting into the building's architecture as it should have been originally.

The original façade flood lighting system not only flattened and concealed the interesting arc shaped feature of the main building, but also made it dull in the evening. A closer look at the façade made it clear that using the balcony's features could highlight the contour of the building by installing a warm white dimmable linear LED fixture. As the construction of the project progressed, the night time lighting make up of the main building was gradually fine-tuned with different dimming levels revealing its original beauty by night.

The cooperation and support from the hotel owner also proved to be cru-

照明设计的故事
Lighting Design Stories

改造前夜晚照片
Before renovation

照明初步概念图
Preliminary lighting concept rendering

cial in other parts of the hotel, where lighting is only applied to some key areas to make those stand out. For the conference center to make the perfect lighting effect in the horizontal cove light above the main entrance, sandblasting was necessary to hide the reflection of the newly installed lighting fixtures by the glossy material. Regarding the logo, unfortunately it was not possible to make it backlit. Nevertheless, it was possible to make it stand out by using projectors on both sides of the conference center.

Shadows are also a part of lighting design in this project. In addition to the arrangement of lights and the control of light sources, colors and angles, the special application of shadows helps create ambiance, highlight surrounding objects and enhance the lighting environment. At one side of the hotel, the reflections of trees dancing on the facade bring the charm and dynamism of the nature to this contemporary still geometric configuration. In the renovation of Shandong Hotel, the lighting designer applied personal understanding of lights to reinterpret the building, creatively polishing the exterior and revealing the building with a brand new look. When night falls, Shandong Hotel stands out from the crowd of buildings with vigor and novelty.

第一章 酒店 | Chapter 1 Hotel
大禹·开元精品酒店 | Dayu Kaiyuan Boutique Hotel

名称：大禹·开元精品酒店
地点：中国绍兴
完工时间：2011 年
占地面积：43 700 平方米
客户：开元房地产开发有限公司
照明设计：英国莱亭迪赛灯光设计合作者事务所－中国分部

Name: Dayu Kaiyuan Boutique Hotel
Location: Shaoxing, China
Year of Completion: 2011
Site Area: 43,700 m²
Client: New Century Real Estate Development Co., Ltd.
Lighting Design: Lighting Design Partnership International (LDPi) China Branch

夜游

绍兴大禹·开元酒店是一家以江南水乡民居风格为主题的高端精品酒店,位于绍兴市东南方向,地处会稽山脚下、毗邻禹陵景区,前身为禹裔后嗣居住地,距今已有四千余年历史。"平和、内敛、精致、儒雅"是开元把原始民居改造成具备江南文人气质酒店的最初目标。酒店在改造中坚持完整地保留禹陵村江南水乡的格局、不损坏一草一木的理念,百余栋民居被低调地改造成客房、茶楼、餐厅、宴会厅等,而古戏台、官河埠头、牌坊、天井、古街、石巷等公共空间更是原封不动。大禹·开元酒店的户外景观照明设计中,最大的特点就是"依景造景,对景成趣",

Night Journey

Situated in southeast of Shaoxing, Dayu Kaiyuan Hotel is a high-end boutique hotel with the features of riverside towns in the south of the Yangze River. By the foot of Kuaiji Mountain, neighboring Dayu Mausoleum, this former residential area of Yu's decents has over 4,000 years of history. Peacefulness, modesty, elaboration and elegance are the final temperaments that the transformation from the ancient residence into a culture hotel aims to achieve. Following the idea of keeping the original structure

第一章 酒店
大禹·开元精品酒店 | Chapter 1　Hotel
Dayu Kaiyuan Boutique Hotel

即利用原建筑及原有景物，运用灯光营造出新的景观。夜晚，坐着乌篷船沿河道抵达客房的途中，墙面的黑白剪影一路缓缓随行，散发着水墨丹青的淡雅气息，在灯光的作用下，斑驳的树影、粼粼的水波、四周的绿树及小桥浑然一体。

在夜幕笼罩下，走在古老的街道上或是坐在乌篷船中，感受着水乡别样的柔美与静谧，错落有致的光线牵引着人们的视线。"以景点为铺垫，以视线点为牵引"是大禹·开元酒店户外景观照明设计的另一个特点。光的动线沿着视线点移动，每走到一个地方驻足而望时，光线总会覆盖到视线所及的范围上。同时设计师捕

of Yuling Village without damaging the local environment, over hundreds of folk houses were quietly renovated into hotel rooms, tea houses, restaurants, banquet halls and etc. Meanwhile, all the public areas such as ancient stages, quays, archways, courtyards, ancient streets and brick lanes remain the same.

The landscaping lighting of Dayu Kaiyuan Hotel makes good use of the building itself and the existing scenery. With the smart application of light-

捉人们在路径观看上的选择，比较哪些景观更具有视觉吸引力，从而进行现场点位的布设，例如戏台、凉亭等视觉点比较多的地方，既最大可能地减少了刻意的人工痕迹，又使灯光在视线的牵引下呈现自然的层次感。

从船埠码头下了船，推开第一扇木门便进入了酒店大堂。大堂原为一户人家的客厅，斑驳的门槛和高悬的横梁显得古朴庄重，大堂门口两侧的迎宾松经过灯效的处理，含蓄地烘托着接待的氛围。客人在大堂取一把旧式门匙，穿越"邻居"的门前屋后，走过小石桥，经过灯笼摇曳的檐廊，就到客房了。有一些客房是带阁楼的屋子，

ing, a brand new night view is created. At night, sitting on Wupeng boats (or black-awning boats) slowly drifting along the river, guests, on their way to the hotel rooms, are able to enjoy the quiet scene and the ink-painting silhouettes on the walls. In the artificial lights, dappled shades, sparkled river, trees and bridges all blend well into a harmonious whole.

As night falls, well-distributed lights guide people's vision. No matter walking on the ancient streets or travelling by water taxis, one will have a

第一章 酒店 | Chapter 1 Hotel
大禹·开元精品酒店 | Dayu Kaiyuan Boutique Hotel

改造后晚上
Night View after Renovation

阁楼上如果没有照明，尤其是阴天，会感觉窗外黑漆漆的。由此，设计师特意用投光灯将屋脊和树植做出光效，展现出渐变的层次感。这样，当客人躺在房间或凭窗遥望时，视线所及之处非常温暖。

在夜晚，最原始、最自然的光就是月光。在这个项目中，为了使夜晚的大禹·开元酒店仍能自然地呈现如诗如画的美景，设计师采用人工照明，营造出仿佛月光投射的自然剪影。用尽量少的人工照明刻画了大禹古村本来的样貌和自然的意境。

大禹·开元酒店作为唯一的酒店入选了由《新旅行》杂志举办的第三届旅行生态大奖，入选的理由是：这不仅是一个酒店，也是一处完美的旅行目的地。

touch of the unique tenderness and quietness of a riverside town. Tourist attractions serve as the leads and viewpoints orientate the way. Streams of light move along tourists attractions and particular scenic views, ensuring brightness is always there in the range of immediate view whichever direction one is looking from. Through a thorough scenery analysis, the most visually attractive landscape contents are highlighted by proper lighting. Where the density of viewpoints increases, efforts are made to minimize artificial lighting while keeping interesting multiple lighting layers.

照明设计的故事
Lighting Design Stories

手绘插图 1
Hand-drawn Illustration I

手绘插图 2
Hand-drawn Illustration II

改造前晚上
Night View before Renovation (highquality picture is not available)

改造后白天
Day View after Renovation

Getting off at the quay, guests enter the lobby which used to be a living room of a folk house. Dappled thresholds and highly hanged cross beams bring out an aura of tradition and elegance. The welcoming pines at both sides of the lobby entrance nail the function of the venue implicitly. With the old-fashioned keys allocated by the receptionist, guests walk past 'neighbors', stone bridges and dim corridors with swaying lanterns and finally arrive at their own rooms. Since some rooms have attics which are originally lacked lighting, spotlights are applied to create lighting effects

第一章 酒店
大禹·开元精品酒店
Chapter 1 Hotel
Dayu Kaiyuan Boutique Hotel

改造后晚上
Night View after Renovation

照明设计的故事
Lighting Design Stories

改造前晚上
Night View before Renovation

on the roof and plants. A stereoscopic feeling then emerges. In this way, no matter one is lying on the bed or looking out by the window, where the vision reaches there is always warmness.

Moonlight is the most natural light in the evening. In this project, to naturally nail the beauty of Dayu Kaiyuan Hotel, artificial lights are installed to create silhouettes that are as if projected by the moon. Reducing the application of artificial lighting accentuates Dayu Village's original style and natural features.

第一章 酒店 | Chapter 1 Hotel
大禹·开元精品酒店 | Dayu Kaiyuan Boutique Hotel

改造后客房窗外景色
Night Scenery outside Hotel Room Window after Renovation

照明设计的故事
Lighting Design Stories

改造后餐厅
The Restaurant after Renovation

Dayu Kaiyuan Hotel is the only hotel to be rewarded the 3rd Travel Ecological Awards by *Voyage*, for it is not just simply a hotel, but a perfect daytime destination and night journey.

第一章 酒店 | Chapter 1 Hotel
绍兴咸亨新天地酒店 | Shaoxing Xianheng Hotel

名称：绍兴咸亨新天地酒店
地点：中国绍兴
完工时间：2010年
总建筑面积：48 004.17 平方米
客户：绍兴咸亨新天地酒店有限公司
室内设计：杭州陈涛室内设计有限公司
　　　　　中国美院风景建筑设计研究院
照明设计：英国莱亭迪赛灯光设计合作者事务所 -中国分部

Name: Shaoxing Xianheng Hotel
Location: Shaoxing, China
Year of Completion: 2010
Gross Floor Area (GFA): 48,004.17 m²
Client: Shaoxing Xianheng Hotel Co., Ltd.
Interior Design: Hangzhou Chentao Interior Design Co., Ltd.
　　　　　　　　The Design Institute of Landscape & Architecture of China Academy of Art
Lighting Design: Lighting Design Partnership International (LDPi) China Branch

献礼中国文学

鲁迅先生的《孔乙己》是20世纪中国文学史上最经典的短篇小说之一。他塑造了孔乙己这位身着长衫、自命清高、迂腐不堪的读书人，给代代中国人留下了深刻的印象。咸亨新天地酒店这个演绎孔乙己落魄一生的地方也因此名扬海内外。"咸亨"二字源于易经坤卦"品物咸亨"。"品物咸亨"意为"万物得以皆美"。咸亨酒店由鲁迅先生的堂叔周仲翔于清光绪二十年创建，几年后关门歇业。1981年为纪念鲁迅先生诞辰100周年重新开业，2007年此酒店被改建成五星级文化主题酒店。时代的变换让昔日孔乙己吃茴香豆的酒铺成为今日游人流连的地方，绍兴老区的

Tribute to Chinese Literature

KONG Yiji written by LU Xun is one of the most classic short novels in Chinese literature in the 20th century. Created by LU Xun, KONG Yiji, an arrogant and pedantic scholar wearing a long gown, has left extremely deep impressions to many Chinese. Xianheng Hotel, the place that witnessed KONG Yiji's difficult life, thus is well-known at home and abroad. Xianheng originated from the Kun Diagram in the Book of Change, meaning everything achieves full development. The hotel was established by

第一章 酒店
绍兴咸亨新天地酒店 | Chapter 1 Hotel
Shaoxing Xianheng Hotel

风貌也在新建咸亨新天地酒店的烘托下显得原汁原味。虽说在配置上,新加入的前区特色休闲购物、后方南北生活馆和客房让现代五星级酒店的水准得到了充分体现,但为了使这样一个新旧交融、传承亦有创新且富有延续性的徽派建筑群在夜晚也能呈现出整体和谐的视觉效果,设计师在开始时就全盘筹划,按场地功能来分析,以照明来区分其功能,但又不喧宾夺主。比如,商业部分相对酒店部分的亮度要高一些,同时也有细节部分的微调,如酒店呈开放状态,出入口非常多,且出入口的位置亮度偏高。

细部的灯光设置基本以简洁为主。在出入口和屋檐上,仅设置了一些必需的功能

LU Xun's uncle ZHOU Zhongxiang in 1894 during the Qing dynasty, but was closed a few years later. In 1981, it was reopened to memorize LU Xun's 100th birthday and later in 2007, it was rebuilt into a five-star cultural theme hotel.

Time has changed. The wine shop where KONG Yiji used to eat fennel-flavored beans, now has transformed into a place where tourists indulge in. The traditional features of Shaoxing's old district are properly foiled with the help of Xianheng Hotel. The newly-added shopping mall, recrea-

灯和环境照明，而就是这种感觉，让人更感惬意。室内设计师在空间中营造了一个独具匠心的视觉焦点，以此展现对中国文学的热衷。踏入酒店大堂，三个装饰吊灯之后，三支巨大的毛笔赫然蘸在长方形砚台上。灯光设计师仅在笔杆处装配少量的灯具，力图以一种含蓄的照明方法突出笔毛。仿佛一切就绪，只等鲁迅先生提笔撰书。

内庭院设有观景台及相连的小桥，稍作烘托，便可以把月光的美感表现出来。八角亭的照明方式运用了中国国画中着墨和留白的处理方式，亭台本身没有特别打光，而内部则设计了一个形似荷叶的倒影般的小型灯光装置，勾勒出如同荷叶的

tional center and hotel rooms are the perfect representation of the standards and quality of a contemporary five-star hotel. This is a place where traditions embrace modernity, where inheritance meets sustainability. To create a harmonious visual effect for such unique Anhui-style complex at night, an overall planning was done at the beginning of the project, analyzing venues according to functions and differentiating them with appropriate lighting. The lighting in the commercial areas of the building is brighter than that of the hotel. However, the lighting of the hotel is fine tuned. Since

第一章 酒店 | Chapter 1 Hotel
绍兴咸亨新天地酒店 | Shaoxing Xianheng Hotel

光影在水上飘浮的效果,一眼望去,周边并不需要光线,它就静谧地存在于那里。生活馆采用比较干净的线型光进行处理,没有太多的光源,主要的电梯厅做了提亮,给人干净、通透的感觉。设计师不希望在已有的完整风格里加一个突兀、明显的装饰品,所以从生活馆一直到前面的商业区都选用了一款简洁的亚力克条形灯,映衬建筑的白色。休闲商业区的灯光处理也很简洁,主要引导人们关注商场的内部,故建筑上没有太多的设置。

咸亨新天地酒店的室内灯光设计效果主要凸显一种比较柔和自然的氛围,让人宾至如归,可以放松地在这里享受时光。设计师采用洁净、透亮的光影来表现客房、

it has many entrances, the level of brightness in doorways is relatively higher than the rest.

Detailed lighting design seeks for simplicity. Some functional lights and ambient lights are installed in the doorways and the roof, creating the sense of comfort and ease. The interior designer created a unique focal point celebrating this unique passion for Chinese literature. When entering the hotel lobby, behind the three decorative pendant lamps, guests can't miss the three massive Chinese paint brushes lying over a black

中西餐厅、宴会厅、大堂等功能空间。虽然酒店的整体建筑是徽派风格，但从内部来看却非常时尚，所以现代风格吊灯点缀在内部空间也不显得突兀。高高的穹顶由悬挂的灯带提点打亮，辅以主应景陈设，如毛笔、竹叶的光影烘托，配合其他落地灯、台灯光度的处理，形成一种明暗的层次变化。

风味餐厅最重视使用者的舒适度，因此主要把灯光集中在餐桌上，其中小吃店面也作了强调。西餐咖啡厅则有所不同，其中大量金属和玻璃元素构造出一种华丽金属的质感，现代却不休闲，相对更加工业化，所以西餐咖啡厅主要运用灰光灯对现代感进行强调。

rectangle ink tray. The lighting designer could only create a subtle design by highlighting the "hair" of the brushes with a few simple lighting fixtures hidden inside the handle, making them ready to be inked under the hand of LU Xun.

Terraces and bridges in the courtyard are connected with each other. With only simple design, the softness, coolness and beauty of the moonlight can be perfectly nailed. The lighting in the octagonal pavilion adopts techniques in Chinese painting, combining inking with blanks that are

绍兴咸亨新天地酒店 | Shaoxing Xianheng Hotel

多功能厅根据需求会有就餐模式、会议模式、放映模式和婚宴模式,所以在灯色转换上比较下功夫,从白色到暖黄色直至红色,灯光设置非常有层次,考虑到有时表演的需要,还特别设置了舞台灯。

夜色渐浓,遥想湖光山色间那些清影白墙,树鸦昏晓中渐渐透出的灯光,慢慢将咸亨新天地勾勒,悄悄将光影腾挪。《孔乙己》中的咸亨酒店,三开间平屋,无窗,昏暗,无隔音墙,曲尺形柜台上放着香醇的黄酒、茴香豆、豆腐干、百叶包、烩鸡蛋等。如今,酒还是那个酒,茴香豆的味道也还在,只是咸亨酒店已经成为了五星级文化风情主题酒店,在灯火阑珊中,续写着它的温馨舒适与历史传承。

intentionally left. No light is specially designed for the pavilion from the outside; however, in the pond of the octagonal pavilion, there is a custom-made lighting in the shape of lotus leaves. Looking from afar, it seems as if the shadows of leaves were silently floating on the water.

Linear lights are used on the façade of the recreational center. With only the necessary quantity of lighting fixtures, the elevator lobby is brightened up. Clarity and transparency are the features that the designer would like to accentuate. Rather than a piece of exaggerated decoration, linear lights

照明设计的故事
Lighting Design Stories

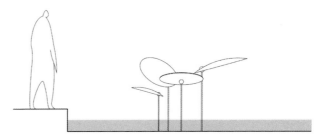

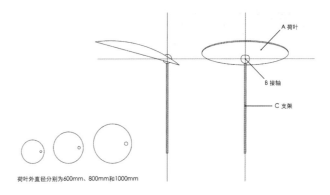

荷叶外直径分别为600mm、800mm和1000mm

are used not only for the recreational center but also for the commercial area to highlight the white color in this complex. The lighting design in shopping mall is also very simple. It is the interior that should be the center of the attention.

The interior lighting design of the Xianheng Hotel is natural and tender, seeking to create a home away from home for those who stay in. Even though this is an Anhui-style complex, transparent lights and clear-cut shadows in hotel rooms, restaurants, banquet halls and lobbies bring out

a contemporary look. Thus modern pendant lights are suited well with the entire design. The dome with hanged strip lights helps nail the main structure of the building. Accompanied with floor lamps and reading lamps, a contrast between brightness and dimness is thus formed.

The level of comfort is essential in restaurants. Lights are mainly centered at dining tables and snack bars; whereas the design of the cafe is modern with a sense of industrialization. Elements such as metal and glass are largely applied to create a feeling of luxury and quality. Therefore, limelight is used to highlight the contemporary features.

The multi-function hall can be arranged into dining hall, conference hall, movie theater or banquet hall according to different requirements. Therefore, efforts are made to the color of lights. From white to warm yellow to red, the color can be adjusted gradually. Besides, stage lights are also installed for the needs of occasional live performances.

As night falls, Xianheng Hotel sits silently. White walls and illusive shadows are faintly visible among mountains and water, dim lights shine through trees and brightness and darkness echo with each other. Xianheng Hotel, in the novel KONG Yiji, is a dark windowless flat with three rooms, without sound insulating walls, an open counter, with Shaoxing wine, fennel-flavored beans, dried tofu, tofu skin rolls with minced pork and stewed eggs lay on the top. However, now in reality, Shaoxing wine and fennel-flavored beans remain the same, but the place has transformed into a five-star culture themed hotel continuing to write its own history and passing on comfort and warmness in the waning light.

第一章 酒店 | Chapter 1 Hotel
宁波开元名都大酒店 | Ningbo New Century Grand Hotel

名称：宁波开元名都大酒店
地点：中国宁波
完工时间：2007 年
总建筑面积：73 148 平方米
客户：开元房地产开发有限公司
室内设计：WATG
建筑设计：WATG
照明设计：英国莱亭迪赛灯光设计合作者事务所－中国分部

Name: Ningbo New Century Grand Hotel
Location: Ningbo, China
Year of Completion: 2007
Gross Floor Area (GFA): 73,148 m²
Client: Kaiyuan Real Estate Development Co., Ltd.
Interior Design: WATG
Architectural Design: WATG
Lighting Design: Lighting Design Partnership International (LDPi) China Branch

Lighting Design Stories

少即是多

宁波开元名都大酒店为五星级豪华酒店,建筑风格呈现代感很强的弧形,位于宁波鄞州路以北,童南路以东,西南与三桥江毗邻。室内和建筑设计由美国知名设计公司WATG完成,是一个运用了较少的光来精确表现酒店各种元素的案例。为了彰显酒店的豪华气派,在室内设计上多采用大型花饰元素,如大堂吧等处带花纹装饰的柱子,多功能厅、宴会厅内各种花形装饰灯,酒店各处的地毯、电梯厅、走廊等处的装饰品。在这纷繁的室内设计中,照明设计师考虑用最简洁的光还原这些花形来凸显夜晚中酒店的华美。先从平面布光着手去分析哪些地方需要照亮

Less Is More

As a 5-star luxury hotel, Ningbo New Century Grand Hotel is a fairly contemporary arch-shaped architecture. It is situated to the north of Yinzhou Road and to the east of Tongnan Road, neighboring Sanqiao River. Designed by WATG, Ningbo New Century Grand Hotel is just a great example of revealing different elements it conveys with less lighting. Large floral ornaments are widely designed in the interior to highlight the hotel's luxury. Those elements are seen on the columns in the lobby

第一章 酒店 | Chapter 1 Hotel
宁波开元名都大酒店 | Ningbo New Century Grand Hotel

而哪些地方不需要，在强调酒店照明具有满足功能用途的情况下，用光使一个空间更有层次感。酒店的主入口彰显豪华气派又不失酒店自身的温馨感，采用了色彩柔和的灯光予以不同层次的照明。在宴会厅照明设计中，考虑到会议、婚宴等不同功能，设计师采用不同色彩为同一个空间营造不同的感觉。

酒店的外立面LED线型灯勾勒了酒店的整体轮廓，立面布满了点光源，正是这些闪光点，带来了整个酒店弧形外立面的明暗韵律。为了在展现夜晚酒店的华丽的同时，不影响客人休息，设计师把LED线型灯条隐藏在阳台上。

在宁波开元名都大酒店的照明设计上，设计师一直坚持用最简洁的灯光去实现最

bar, the decorative lights in multi-functional rooms and banquet halls, the carpets, elevators and the decorations in corridors. Only the simplest lighting is applied to nail the gorgeousness of the colorful but rich interior design. In the first step, the analysis of the lighting distribution is done to accentuate the layering of the space while satisfying the functional lighting. The main entrance is with soft-colored and layered lighting to show the scale and welcoming of this hotel. In banquet halls, considering the variety of the events such as conferences and weddings to be held there, differ-

照明设计的故事
Lighting Design Stories

好的效果。如玄关和日本餐厅，将需要用光的地方照亮即可。同时，我们尽量选用极少种类的灯具去满足酒店的照明需求，希望这种做法给酒店日后的维护带来方便。

ent colors of lights are applied so that the space is able to satisfy different requirements.

The facade uses linear LED to portray the outline of the architecture and wall lamps are installed on balconies, which create the interactions of lights and shades at night. Those LED wall lamps are hidden in balconies, making it possible to show the slenderness of the hotel without any disturbance to the guests.

第一章 酒店
宁波开元名都大酒店 | Chapter 1 Hotel — Ningbo New Century Grand Hotel

Lighting Design Stories

The simplest lighting is used in the Ningbo New Century Grand Hotel to achieve the best effects. In the entrance hall and the Japanese restaurant, only those necessary places are lighted up. Additionally, the types of lighting devices installed are reduced to the minimum to reduce the difficulties in the future maintenance.

第一章 酒店 | Chapter 1 Hotel
北京广西大厦酒店 | Beijing Guangxi Hotel

灯光改造前
Before lighting renovation

改造前夜晚照片
Before renovation

照明初步概念图
Preliminary lighting concept rendering

灯光改造后
After the lighting renovation

名称：北京广西大厦酒店
地点：中国北京
完工时间：2007 年
客户：北京广西大厦酒店
照明设计：英国莱亭迪赛灯光设计合作者事务所－中国分部

Name: Beijing Guangxi Hotel
Location: Beijing, China
Year of Completion: 2007
Client: Beijing Guangxi Hotel
Lighting Design: Lighting Design Partnership International (LDPi) China Branch

点亮广西文化

北京广西大厦是一座集住宿、餐饮、会议、康乐于一体的四星级商务酒店，位于北京东三环，毗邻潘家园古玩城，交通十分便利。此酒店修建于1999年，2006年重新装修改造，LDPi 照明团队加入工作。

改造前，广西大厦的外立面只是用白色投光灯大面积地打亮，从远处看过去白白的一片，乏味且毫无层次感，同时这种大面积投光，也照亮了客房，影响了入住客人的休息。雨篷和大堂入口等处很暗，当客人到达酒店时，很难找到入口，带来很多不便。

照明设计师对这个十年前建造的旧建筑详细地调研了一翻。大厦主楼是以桂林象鼻山为灵感建造的，同时融合了壮族民居吊角楼的形式，与裙楼、公寓楼高矮错落、上下呼应，形成了一个小型的建筑群落，并且把广西民族特色与现代设计风格巧

Switch on Guangxi Culture

Situated by the side of the Beijing East 3rd Ring Road, Beijing Guangxi Hotel is a 4-star business hotel with accommodations, restaurants and bars as well as facilities for conference and entertainment. Neighboring the Panjiayuan antique market, the hotel is conveniently located near public transportation. It was built in 1999 and was renovated in 2006 at which point LDPi participated in the project for the lighting design renovation.

Before the renovation, only a very simple lighting was used projecting cool white light on the façade. Not only did the building become flat without contrast, but also it affected guests' privacy. Furthermore, one could miss the main entrance as it was left in the dark, bringing inconvenience when

第一章 酒店 | Chapter 1 Hotel
北京广西大厦酒店 | Beijing Guangxi Hotel

照明初步概念图
Preliminary lighting concept rendering

妙融合,在大厦的外立面等处都能见到各种流行于桂林少数民族区的元素符号或是图腾。

LDPi 照明设计师将亮点锁定在这些源自广西古老的图腾等具有个性和特色的主要元素上,在建筑入口处、酒店标志、外立面上的蓝色琉璃瓦等处做了布光。在主入口增加了功能性照明,并采用 3 000 K 的暖黄光营造温馨的迎宾氛围。设计师通过修改原有的灯具,使用藏在树里的节点灯、往下照明的节能灯、往上照明的六度角的投光灯等多种灯具,来实现大厦的照明改造。

作为十年前的一个并不具有太多影响力的老项目,以今天日新月异的视角来审视,白天依旧没有任何变化,夜色中的广西大厦却具有视觉品位,毫不落伍。这要归功于 LDPi 照明设计师用光表达了建筑的理念,注解了广西文化,用明暗错落的韵律还原了其本质。虽然它不如周围建筑那样明亮,但给人以舒适感。

guests arrived at the hotel.

A careful study was conducted to this old building constructed more than a decade. The design of the main building originates from the Elephant Trunk Hill in Guilin. Adopting the features of the folk houses of Zhuang ethnic minority, it forms a small complex together with the podium and the apartment building, harmoniously combined with the contemporary design, unique symbols or totems on the facades and other places.

The hotel's lighting is centered at the folk elements such as the ancient totems from Guangxi. The lighting design mainly focuses on the entrance, hotel logo and glazed blue tiles to better showcase those features. Functional lighting is also added with 3,000 K warm white light under the

照明初步概念图
Preliminary lighting concept rendering

entrance's canopy to create a welcoming spirit. The renovation of Beijing Guangxi Hotel's lighting is realized by modifying the original lighting and using spot lights, energy saving down lights and 6° projectors.

Beijing Guangxi Hotel, despite has a history of over a decade, looks the same during daytime, is still charming and up-to-date when the night falls where the contrast of brightness and darkness perfectly reveals and interprets the concept and essence of the architecture and Guangxi culture, making it a humble and comfortable building.

第一章 酒店 | Chapter 1 Hotel
南昌洗药湖避暑山庄 | Nanchang Green Lake Resort

名称：南昌洗药湖避暑山庄
地点：中国南昌
完工时间：2012 年
总建筑面积：16 100 平方米
客户：南昌市政公用投资控股集团
建筑设计：筑博设计
照明设计：英国莱亭迪赛灯光设计合作者事务所－中国分部

Name: Nanchang Green Lake Resort
Location: Nanchang, China
Year of Completion: 2012
Gross Floor Area (GFA): 16,100 m²
Client: Nanchang Municipal Public Investment Holding Group
Architectural Design: ZHUBO DESIGN
Lighting Design: Lighting Design Partnership International (LDPi) China Branch

水雾飞鸟

南昌洗药湖避暑山庄位于南昌市西郊15公里处的梅岭，西邻鄱阳湖，北与庐山对峙，自然风景独特。风景区内，距放生池北700米处，有一组自上而下依次排列的三级水池，据说这就是当年葛玄清洗草药的"洗药湖"。依山而建的避暑山庄是由一个衰败闲置了很久的招待所改造扩建而成的五星级酒店。无论从景观到建筑主体，还是室内装饰，避暑山庄都延续了统一的设计定位，即与自然和谐的东方气质。改建过程中仍保留了原建筑结构，并加入了一些东南亚建筑风格的元素，比如大尺寸的挑檐、大量的木材和米黄色调的石材等，由此营造出粗糙的质感，

A Bird in the Mist

Situated in the west suburb in capital of Jiangxi province, about 15 km away from Meiling, Nanchang Green Lake Resort, with its unique scenery, neighbors Poyang Lake in the west and opposites Lushan Mountain in the north. Within the resort, 700 m to the north of the Free Life Pond, lie 3 pools which are said to be the Xiyao Lake where GE Xuan used to wash medical herbs. An idle guesthouse, which was built along the mountain, now has been reconstructed and refurbished to a five-star hotel. This

第一章 酒店
南昌洗药湖避暑山庄 | Chapter 1 Hotel
Nanchang Green Lake Resort

给人一种原始朴素的感觉，好像建筑是融在浑然天成的山水林木里的景色。

梅岭相对地势较低，绿林环抱，拥有常年水雾笼罩的气候特点。一天中的大部分时间都是雾蒙蒙的，水雾会把建筑笼罩起来，所以，身在远处，只能隐约看见建筑的朦胧轮廓，而朝不同方向伸展的挑檐给人一种大鹏展翅的具象联想。

夜晚风景区光线较暗，所以在照明设计上，设计师考虑到景区的水雾特点和原始

resort from the exterior to the interior perfectly reflects the features of oriental culture. The original structure is preserved, while adding some elements, such as cornice, wood and beige stones of Southeast Asia. The architecture, full of nature and authenticity, thus fits wonderfully the local natural scenery.

的建筑风格，选择了低亮度、较暗的光线，让建筑与景观保持较高的契合度。与此同时，照明突出强调顶部挑檐，使建筑的顶部隐隐约约地显露出来。建筑的屋檐和屋顶用光加以强化，赋予其动感，但是考虑到客房会有一些室内的光透出来，建筑底部则不做灯光的处理，由此来保持其自然静谧的本色。为了与周围自然环境保持统一，在主路口的大堂处，只做了少量的灯光处理，光融于自然，减少了

Being at relatively low altitude, Meiling is surrounded by forests and thus is affected by mist enshrouding the architecture frequently during the day. Looking from afar, one could only make out the shape of the roofs that stretch out to different directions, as if a great hawk spreading out its wings.

第一章 酒店 | Chapter 1 Hotel
南昌洗药湖避暑山庄 | Nanchang Green Lake Resort

照明初步概念图
Preliminary lighting concept rendering

人工痕迹。另外，所选灯的功率也都比较低，基本是20瓦或35瓦的卤素灯泡以及功率更小的LED，节省了能耗成本。

由于某些原因，景观照明是由景观设计师来完成的。如果由LDPi对景观部分照明重新设计，我们将考虑不直接打亮绿植，而是利用安装在池塘中的暖白线型灯而非点光源来制造光与影的意趣，同时展现出石头的自然纹理之美。

The scenic area is relatively dark at night; therefore considering the mist and the style of architecture, lighting fixtures with lower brightness and dim lights are applied to integrate the building with its surroundings. Meanwhile, the roof top, being accentuated by the lighting, is faintly perceptible in its outlines. The eaves and ceiling are stressed by beams of

如此的照明设计与国画中的留白有着异曲同工之妙，提升整体美感的同时给人可以呼吸的空间。

lights, which bring them to life; by contrast, there are no lights installed at the bottom of the building, but the authenticity and the quietness of the nature is largely preserved with lights glowing from the guestrooms. At the lobby entrance, only minus lighting is applied to compromise with the natural environment. Limited artificial lighting integrates with the nature.

第一章 酒店
南昌洗药湖避暑山庄
Chapter 1 Hotel | Nanchang Green Lake Resort

夜间灯光效果图
Night time lighting rendering

Lighting Design Stories

建筑效果图
Original architectural rendering

Halogen bulbs (20 w or 35 w) and LED with lower wattage are selected to reduce the cost and save energy.

The landscape lighting was designed by the landscape designer. Nevertheless, if LDPi were to redesign the landscape lighting, it would also keep trees in the dark but create a contrasting game of shadow with warm white linear lights that would be used as underwater fixtures installed in the pond instead of spot lights so as to reveal the beautiful texture of the stone forming the several terraces.

Just like leaving some spaces intentionally blank in Chinese painting, such lighting design would beautify the entire environment and create comfort.

第一章 酒店
苏州新城花园酒店 | Chapter 1 Hotel | Suzhou New City Garden Hotel

照明初步概念图
Preliminary lighting concept rendering

名称：苏州新城花园酒店
地点：中国苏州
完工时间：2008 年
总建筑面积：46 000 平方米
客户：苏州新城花园酒店
室内设计：苏州金螳螂建筑装饰股份有限公司
照明设计：英国莱亭迪赛灯光设计合作者事务所－中国分部

Name: Suzhou New City Garden Hotel
Location: Suzhou, China
Year of Completion: 2008
Gross Floor Area (GFA): 46,000 m^2
Client: Suzhou New City Garden Hotel
Interior Design: Suzhou Gold Mantis Construction Decoration Co., Ltd.
Lighting Design: Lighting Design Partnership International (LDPi) China Branch

第一章 酒店 | Chapter 1 Hotel
苏州新城花园酒店 | Suzhou New City Garden Hotel

飘带

苏州新城花园酒店地理位置优越，坐落于东方水城苏州古老的京杭大运河畔，太湖湿地景区环抱于西，近邻虎丘山、寒山寺等名胜古迹和"世遗"园林，同时地处繁华的商贸金融中心，西行500米即是著名的"商业街"，距古城区中心仅有三公里，是舒适便捷的商务居所。酒店的室内设计，彰显古韵今风相融合的柔美气质，展现了江南水畔花园般的温婉雅致。核心区利用灯光打造了富有动感的飘带设计；配合不同季节主题的中庭景观，照明色彩随之变化，这些特点都给人以清新自然的视觉享受。

Ribbon

With a perfect location right in the heart of the commercial center of the city, Suzhou New City Garden Hotel lies by the side of the Beijing-Hangzhou Grand Canal, with Taihu Lake Wetland Park to its west, neighboring Tiger Hill, Hanshan Temple and World Heritage Gardens. Being 500 m away to the Down Town Street and 3 km to the ancient city area, the hotel is a comfortable and convenient place to stay. Its interior design conveys both traditional and contemporary elements, showing the ele-

照明设计的故事
Lighting Design Stories

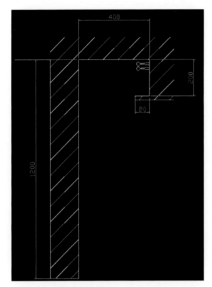

大堂天花飘带照明节点示意：
The schematic diagram of the ribbons' lighting node in the lobby

照明初步概念图
Preliminary lighting concept rendering

酒店别出心裁的灯光设计表现为流线型的飘带。为了将飘逸感充分展现出来，设计师在设计之初进行了尺寸为一比一的模拟实验，对于飘带开孔的大小，也做了多次实际模拟，为了营造如风的飘逸效果使用了PVC材料。飘带的中空及变幻的线条感，通过灯光照明，使得弧形灯带的流动感更加生动与丰富。大堂核心区域大面积米色基调的地面穿插设计了一些淡棕色的曲线线条，与上空飘带相呼映。白天，酒店80%至90%的灯具（包括装饰灯）都是打开的，考虑到太阳光，固定的射灯相对亮度会降低一些。所以，在自然采光及灯光都非常充足的白天，大堂上空发光的飘带会在餐厅等局部暗色调但质地光洁的地板上投射出舞动的倒影。

gance and beauty of the city in the southern Yangtze River. The lights in the main area form a dynamic ribbon, the color of which changes according to the views in the atrium. Such design brings a refreshed and natural visual enjoyment.

The core of the out-of-ordinary lighting design in this hotel is the glowing ribbons. In order to show their elegance and lightness, a full-scale experiment was carried out before the installation. Besides, sizes of holes on the ribbons were also tested. To portray the lightness of the ribbons,

第一章 酒店
苏州新城花园酒店 | Chapter 1 Hotel Suzhou New City Garden Hotel

泳池天花板的灯光设计也来源于飘带的理念,飞舞的飘带挂在天花板顶栩栩如生,配合场景需要照明设计了两种模式:淡蓝色和暖白,分别表现夜晚和白天。

苏州新城花园酒店的主楼为圆柱式建筑,是中国第一家拥有五个中庭花园的酒店,酒店东翼拥有5个挑高式中庭,分别定位为"春""夏""秋""冬""樱"五个主题风格,不同主题是根据各个中庭的特点来进行特别设计的,比如和式房和女宾房就分别设在了"樱"和"夏"的中庭里。照明设计为了突出中庭花园的特色,将LED等灯具嵌在云石等材料中,根据时间的不同营造出不同色彩的变幻效果,五彩缤纷的零星花瓣也可以根据不同的季节来变化色彩。

acrylic panels were used. Through lighting, the hollowness and changeable lines of the ribbons are stressed, increasing their floating nature. The cream-colored floor in the lobby has light brown curves intertwined with each other echoing with the ribbons above. During the day, 80% to 90% lights including decorative lighting in the hotel are turned on; therefore considering sunlight, the brightness of spotlights is reduced. Hence, with sufficient natural light and artificial lighting in the day time, the lobby, the restaurant and other areas with dark but smooth floor have the reflection

夜间灯光效果图
Night time lighting rendering

作为商务酒店，在室内灯光设计上，低调的奢华与静怡是最常规的风格要求，同时苏州新城花园酒店运用了大量丰富的灯光色彩和创意，细节处显示了照明设计师的巧思。

of the glowing ribbons on the ceiling. The lighting design of the roof of the swimming pool also originates from the ribbon. Floating freely, light curve bands hanging upon the ceiling are designed with 2 modes, which are warm white and light blue, to reflect the day and night respectively.

Suzhou New City Garden Hotel, with its main building being cylindrical, is the first hotel in China that has 5 gardens in its 5 atriums. The east wing of the hotel has 5 high-raised atriums, which are decorated into 5 themes, namely spring, summer, autumn, winter and cherry. Japanese-style

第一章 酒店 | Chapter 1 Hotel
苏州新城花园酒店 | Suzhou New City Garden Hotel

Lighting Design Stories

rooms and the female guest rooms are located in the cherry atrium and summer atrium respectively. The lighting design there aims to accentuate the features of the atrium, uses LED lights and embeds them into building materials. The color of lights changes according to the time and the pastels on the wall also change color based on different seasons.

As a commercial hotel, the interior lighting design follows the style of preserved luxury and quietness. At the same time, colors and creativity are also largely implied in Suzhou New City Garden Hotel, reflecting the lighting designer's ingenuity.

商业综合体
Retail and Mixed Use
Chapter 第二章
02

宁波和义大道购物中心
成都来福士广场
北京金隅万科广场
广州太阳新天地购物中心
北京来福士中心
成都凯丹广场
北京冠城名敦道综合体

Ningbo Heyi Avenue Shopping Center
Chengdu CapitaLand Raffles City
Beijing Jinyu Vanke Plaza
Guangzhou Happy Valley
Beijing CapitaLand Raffles City
Chengdu GTC Galleria
Beijing Midtown

Chapter 2 Retail and Mixed Use
Ningbo Heyi Avenue Shopping Center

名称：宁波和义大道购物中心	**Name:** Ningbo Heyi Avenue Shopping Center
地点：中国宁波	**Location:** Ningbo, China
完工时间：2009 年	**Year of Completion:** 2009
占地面积：58 800 平方米	**Site Area:** 58,800 m²
总建筑面积：124 063 平方米	**Gross Floor Area (GFA):** 124,063 m²
客户：宁波市海城投资开发有限公司	**Client:** Ningbo Haicheng Investment Development Co., Ltd.
建筑设计：Denton Corker Marshall（DCM Studios）	**Architectural Design:** Denton Corker Marshall (DCM Studios)
照明设计：英国莱亭迪赛灯光设计合作者事务所－中国分部	**Lighting Design:** Lighting Design Partnership International (LDPi) China Branch

平面、直线和圆弧

在 LDPi 的商业照明设计项目中，宁波和义大道是一个很有特色空间的项目。在项目最初的照明设计研究阶段，设计师发觉购物中心的建筑结构别具匠心，兼顾了亚洲与欧洲的特色。建筑设计不仅汲取了水稻梯田的设计理念，同时也深受西方"平面、直线和圆弧"立体派主义的影响。

后来，当我们看整个建筑的平面图时又有了惊人的发现：整个建筑群和毕加索

Planes, Lines and Arcs

The Ningbo Heyi Avenue shopping center holds a special place among LDPi's commercial lighting design projects. Indeed, during the preliminary lighting design study phase of the project, the architectural features were found to be a striking mix of Asian and European styles. Not only the original architectural design was influenced by Asian rice fields' terraces,

第二章 商业综合体
宁波和义大道购物中心
Chapter 2　Retail and Mixed Use
Ningbo Heyi Avenue Shopping Center

1921年的画作"三位音乐家"极其相似。一切看上去都那么贴合：人数、形状、头部，特别是Pierrot的右手臂！

立面照明充分考虑了业主以及路易威登的要求，采用线形设计，干净优雅地体现出了宁波和义大道上的平面、直线和圆弧元素，高端大气的购物中心从周围庞杂的建筑中脱颖而出。该建筑群中的3个椭圆和2个长方形的玻璃天井也以类似的手法处理。

but also it was found to have a strong cubism influence with 'planes, lines and arcs'.

The later influence became amazingly striking to the project team when looking at the overall architectural plan. It arguably resembled the 'Three Musicians' painted by Picasso in 1921. All seemed to match: body count, shapes, heads and especially the right arm of the Pierrot.

照明设计的故事
Lighting Design Stories

面朝河水的弧形餐厅的外立面是整个购物中心里唯一一处被施以颜色的地方。设计师分别使用了小模型灯光和真实比例的模型灯光来进行实验，保证设计效果的实现。

从纽约出差归来，设计师带了一幅来自现代艺术博物馆的影印缩小版《三位音乐家》，置于LDPi办公室中。

The general façade lighting, in alignment with the requests from the property owner and Louis Vuitton, was linearly designed to cleanly and elegantly highlighting the 'planes, lines and arcs' of the Ningbo Heyi Avenue, thus distinguishing itself with a high end status from other disordered surrounding properties. Following this concept, all 3 oval and 2 rectangular shaped atrium glass boxes with patterns were treated in the same way.

照明设计的故事
Lighting Design Stories

照明初步概念图
Preliminary lighting concept rendering

The only location where color was intentionally used is on the façade of the independent arc-shaped restaurants facing the river bank for which a small scale and real scale mockups were necessary to ensure the designed effect could be successfully achieved.

Following a business trip to the New York, Museum of Modern Art, a small printed copy of the 'Three Musicians' was brought back to LDPi's office.

第二章 商业综合体 | Chapter 2 Retail and Mixed Use
成都来福士广场 | Chengdu CapitaLand Raffles City

照明设计的故事
Lighting Design Stories

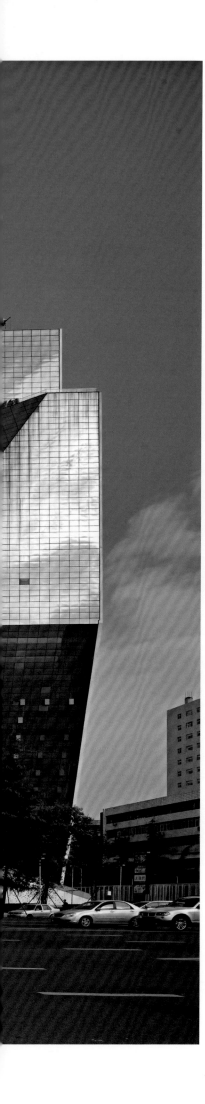

名称：成都来福士广场
地点：中国成都
完工时间：2013 年
占地面积：32 571.78 平方米
总建筑面积：308 939 平方米
客户：成都来福士实业有限公司
办公区室内设计：Draugtzman (DZ Hong Kong)
酒店室内设计：BBG-BBGM(BBG USA)
零售区室内设计：The Buchan Group (TBG Australia)
建筑设计：Steven Holl Architects (SHA)
景观设计：Steven Holl Architects (SHA)
发光雕塑设计：Lebbeus Woods and Christophe a. Kumpusch (LWCK) SHA，LDPi
照明设计：L'Observatoire International (OBS)
　　　　　英国莱亭迪赛灯光设计合作者事务所 –中国分部

Name: Chengdu CapitaLand Raffles City
Location: Chengdu, China
Year of Completion: 2013
Site Area: 32,571.78 m²
Gross Floor Area (GFA): 308,939 m²
Client: Chengdu Raffles City Co., Ltd.
Office Interior Design: Draugtzman (DZ Hong Kong)
Hotel Interior Design: BBG-BBGM (BBG USA)
Retail Interior Design: The Buchan Group (TBG Australia)
Architectural Design: Steven Holl Architects (SHA)
Landscape Design: Steven Holl Architects (SHA)
Light Pavilion Design: Lebbeus Woods and Christophe a. Kumpusch, LWCK, SHA, LDPi
Lighting Design: L'Observatoire International (OBS)
　　　　　Lighting Design Partnership International (LDPi) China Branch

设计公司参与环节 Designers involved per phase	方案设计 Schematic Design	扩初设计 Preliminary Design Development	深化设计至现场指导 Design Development until Site Administration
外立面 Façade	SHA \ OBS	SHA \ OBS	SHA \ LDPi
景观 Landscape	SHA \ OBS	SHA \ OBS	SHA \ LDPi
办公室内 Office Interior	DZ \ LDPi	DZ \ LDPi	DZ \ LDPi
酒店室内 Hotel Interior	BBG	BBG \ LDPi	BBG \ LDPi
商业室内 Retail Interior	TBG	TBG \ LDPi	TBG \ LDPi
发光雕塑 Light Pavilion	LBWCK	SHA \ LDPi	SHA \ LDPi

第二章 商业综合体 | Chapter 2 Retail and Mixed Use
成都来福士广场 | Chengdu CapitaLand Raffles City

流畅

成都来福士广场位于人民南路四段东端，四川省体育馆的正对面。这片荣获绿色能源与环境设计先锋奖金级认证的建筑群内部与公共交通完美结合，总楼层面积达240 928平方米。此外，来福士还包括甲级办公室、服务式公寓、购物中心以及雅诗阁酒店。

Fluid

The CapitaLand's Chengdu Raffles City is located opposite of the Sichuan Gymnasium at the beginning of the east side of the 4th section of the South Renmin Road. It is a LEED gold certified building, fully interconnected with public transportation, providing access to this mixed used

照明设计的故事
Lighting Design Stories

Steven Holl Architects (SHA) 设计的成都来福士广场充分运用了自然元素，通透的建筑与周边的光线和水呼应得恰到好处。自然光不仅能够通过楼体中大面积的开口、断面和入口投射进来，同时还可以穿过水景照入，从而在白天与人工照明相辅相成。相反，在夜幕降临后，内部的人工照明则会由楼体内部散射到外部。此外，在3号楼低层，照明灯具被精心设计成发光雕塑，并在其周围环绕着

project with a gross floor area of 240,928 m². It includes an office of class A, serviced residences, shopping mall and an Ascott Hotel.
Natural elements were fully considered in the design of the project. By designing a 'porous' architecture, Steven Holl Architects (SHA) enables an interaction of light and water with the public.

第二章 商业综合体 | Chapter 2　Retail and Mixed Use
成都来福士广场 | Chengdu CapitaLand Raffles City

玻璃。这个设计的用时长，复杂程度也高，灯具内部安装的 LED 灯带就长约 3 000 米。为确保照明色彩能够与 Lebbeus Woods and Christophe a. Kumpusch(LWCK) 的设计理念吻合，SHA 和 LDPi 密切合作并在包含红、绿、蓝 LED 光点的发光体上又添加了白色的 LED 光点。

另外，照明设计还很注重与水元素的呼应和结合。18 个分别安装在 1、3、5 号

Natural light penetrates into the interior space not only through the numerous openings, cuts and entrance, but also through the waterscapes, thus complementing the artificial light used during daytime. But at night a reverse exchange function comes into effect where artificial light glows from within and escapes outward.

楼上的投光灯打在三个水景上,在建筑内部三个立面和商场中形成令人惊艳的水纹。在1号楼屋顶花园的上方安置的水下灯有意向上打光,在顶棚上形成了很小型的花纹。不过,这种景象的形成所需的先决条件是水面保持相对平静。

在照明设计上运用水元素的灵感源于杜甫的诗作,楼梯内部的三处水景分别代表西陵峡、巫峡和瞿塘峡,每座"峡"分别根据其自身特征进行了照明设计。五层

Furthermore, light is also celebrated as a dedicated pavilion glowing at night in the lower part of tower 3 and surrounded by reflecting glass. Its lengthiness and complexity are manifold. Indeed, there is about 3,000 m of LED strip lights installed within. And in order to provide accurate color output to match the design intent from Lebbeus Woods and Christophe a.

楼高的圆形中厅代表了西陵峡的高耸陡峭,中厅里的单双头灯布光分别体现了峡谷的唯一性和两面性。4 000 K荧光板条置于可伸缩的材料上形成白色雾霭效果,则恰似巫峡的雾气缭绕。瞿塘峡的温暖不仅通过建材的花纹和色彩来体现,还通过安装在中厅上方的3 000 K暖白色筒灯完美诠释。

成都来福士广场处处体现出了流畅这一特点:项目进行中,不同的设计团队与凯

Kumpusch (LWCK), SHA and LDPi cooperated closely before choosing to add a dedicated white LED dot to the standard red, green and blue LED dots on the linear LED fixture.

Last but not least, when combined with water, amazing wave patterns appear by shooting lights onto the three waterscapes from 18 projectors

第二章 商业综合体 | Chapter 2 Retail and Mixed Use
成都来福士广场 | Chengdu CapitaLand Raffles City

德集团的合作可谓是顺畅；同时自然与人、自然元素与人工元素之间的相互呼应和结合所产生的惊艳效果也流畅地贯穿在整个项目当中。

installed on tower 1, 3 and 5. These patterns materialize both on three inner façades and inside the shopping mall; on the top of a roof garden on tower 1, underwater lights are intended to shoot upwards and create a smaller pattern on the ceiling. The only precondition is that the water should not be strongly agitated.

Lighting Design Stories

Water, with an inspiration from DU Fu's poetic works, connects the exterior landscape and the interior with three waterscapes representing the Xiling Gorge, Wu Gorge, and Qutang Gorge. Each gorge received a particular treatment based on its characteristics. The height and steepness of the real Xiling Gorge is represented by the five-storey high round shaped atrium. This duality and singularity is also captured by applying a lighting layout which uses double and single head lights around the atrium. The mist and cloudiness of the Wu Gorge is revealed through a white glowing cloud-like ceiling where 4,000 K fluorescent battens are installed above the stretched material. The warmness of the Qutang Gorge is perceived not only through the colors and patterns of the material used, but also through the 3,000 K warm white down lights installed under this special shaped atrium.

For this project, fluidity is revealed and incorporated in various aspects. It is in the design of the project with cooperation between designers themselves and with CapitaLand, in the considerations between nature and mankind, in the interaction and output between natural and artificial elements.

第二章 商业综合体 | Chapter 2 Retail and Mixed Use
北京金隅万科广场 | Beijing Jinyu Vanke Plaza

图片由万科提供
Picture courtesy of Vanke

名称：北京金隅万科广场	**Name:** Beijing Jinyu Vanke Plaza
地点：中国北京	**Location:** Beijing, China
完工时间：2013 年	**Year of Completion:** 2013
占地面积：23 000 平方米	**Site Area:** 23,000 m²
总建筑面积：140 000 平方米	**Gross Floor Area (GFA):** 140,000 m²
客户：北京金隅万科房地产开发有限公司	**Client:** Beijing Jinyu Vanke Real Estate Development Co., Ltd.
室内设计：BENOY	**Interior Design:** BENOY
建筑设计：Woods Bagot	**Architectural Design:** Woods Bagot
景观设计：AECOM、麦田景观设计院	**Landscape Design:** AECOM, Maitian Landscape Design Institute
照明设计：英国莱亭迪赛灯光设计合作者事务所-中国分部	**Lighting Design:** Lighting Design Partnership International (LDPi) China Branch

照明设计的故事
Lighting Design Stories

梦幻宝盒

北京金隅万科广场项目地处北京昌平城区的核心地段南环路和创新路交界处，不仅交通便利，更荟萃时尚百货、生活超市、娱乐餐饮及其他服务于一体，致力营造都市休闲娱乐社交的综合商业氛围。

照明设计师在研究了建筑设计结构后，对玻璃盒体和穿孔式金属板防水墙采用不同的设

Magic Treasure Box

The Beijing Jinyu Vanke Plaza is located in the core intersection of Nanhuan Road and Chuangxin Road of Changping District. This project, with convenient transportation, has a department store, supermarket, entertainment facilities, restaurants and other daily life services aiming to

第二章 商业综合体 | Chapter 2 Retail and Mixed Use
北京金隅万科广场 | Beijing Jinyu Vanke Plaza

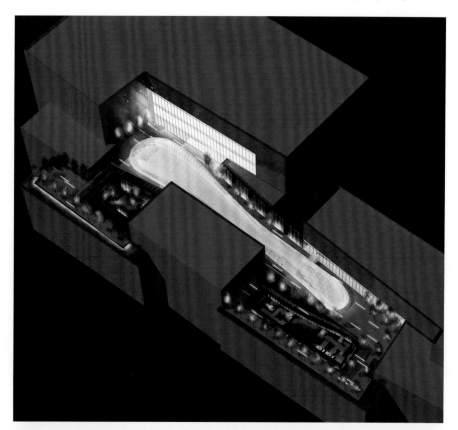

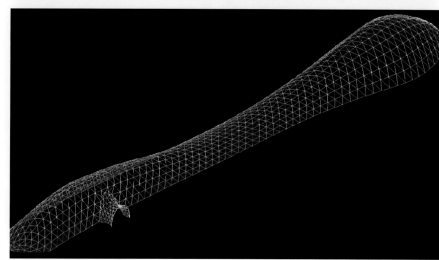

计手法但又保证整体立面和谐统一。在 Wood Bagots 设计的玻璃盒里运用了两种 LED 灯具。幕墙后的树枝状弧形斜肋处暗藏白色 5 500 K 左右的小功率户外 LED 线型灯，光投射在建筑材料上，间接反射的光线较为柔和，在建筑立面上形成弧形光带。采用大功率的户外 RGB 动态可变色线型 LED 上下投光灯打亮，并被有效地隐藏在每一层马道内，增加了立面雕塑质感的同时也使通透的玻璃盒体整体颜色交替变化，呈现缤纷夺目

create a new and contemporary lifestyle.

After examining the structure of the building, the team decided to apply different lighting methods to glass box and perforated metal panel water-proofed curtain wall while achieving an integral overall effect. In the glass box designed by Woods Bagots, two types of LEDs were used. Low-

第二章 商业综合体 | Chapter 2 Retail and Mixed Use
北京金隅万科广场 | Beijing Jinyu Vanke Plaza

的视觉效果，带来梦幻艺术的气息。

为了不喧宾夺主，在金属幕墙的方形结构下方隐藏放置了单一3 000 K暖白色户外LED条形灯，间接反射内透的背光使方形结构在整体立面形成透亮效果，呼应了方形玻璃盒体的晶莹剔透之感。大小不一的方形结构在整体立面上形成了飘带状排布，好似无意中撒上的水晶颗粒，增加了建筑体的动感韵律。光的反射使得整

voltage 5,500 K white outdoor LED strips were hidden under the arc-shaped pattern behind the glass curtain wall, allowing the indirect reflected lights to form a soft arc-shaped light ribbon. High-voltage outdoor RGB linear LED projectors were also hiddenly installed in beams to enhance the texture of the facades while enabling a color changing in the glass box. A

体建筑愈加透明，与玻璃盒交相呼应。建筑入口处广场沿用了方形结构理念，采用5 500 K白色方形户外LED地砖灯营造趣味迎宾的氛围。

屋顶花园作为休闲的休息区，亮度相对较低，打造安静氛围的同时也很好地与自然的月光完美融合。3 000 K暖白LED线型埋地灯装饰和指引道路，同时线型LED投光灯用来呈现座椅的造型。

dazzling visual effect was thus created.

Behind the square-shaped windows with different dimensions on the golden metal curtain wall, 3,000 K warm white outdoor LED lights were used, thus creating another crystal-like ribbon scattered around the façade and enhancing the vitality of the architecture. The reflection of lights makes the

第二章 商业综合体 | Chapter 2 Retail and Mixed Use
北京金隅万科广场 | Beijing Jinyu Vanke Plaza

屋顶花园的主要特色是通过一个特别形状的穹顶天花连接到室内,让人们可以近距离感受色彩变化时的梦幻感和银河的美景。采用 RGB 可变色 LED 线型投光灯沿室内穹顶底部布设一周,呼应玻璃盒体交互变色。

这种梦幻感同样被带到了室内,室内照明除了用光凸显天花板多种图形设计之外,主要是用光纤在天花板上模拟星空,并用光勾勒其他地方天花的不同图形。走廊、

structure even more transparent, which is constant with the translucent feature of the glass box. The square-shape was also used at the entrance where 5,500 K white outdoor LED light bricks were installed to create an interesting and welcoming feeling in connection with the golden façade. Treated as a resting area, the roof garden was kept dimmed to create a

照明设计的故事
Lighting Design Stories

电梯厅、卫生间这些空间兼具了功能照明和装饰照明。
银河带状玻璃穹顶和建筑玻璃盒体在光色的变化中形成了梦幻宝盒,给整个建筑带来了活力和梦幻色彩。如此设计使金隅万科广场不再只是一个购物的地点,更为人们休闲、娱乐和商务社交营造了有趣的艺术空间和氛围。

quiet ambiance allowing people to enjoy the natural environment beautified by the moonlight. There, people are guided through decorative and functional ground recessed 3,000 K warm white LED strips; while benches were also lit with linear LEDs installed underneath the seating area.
But the main feature of the roof garden is an organic shape skylight con-

necting it to the interior space. It offers people a great opportunity to closely experience the joy of a dream and surprising view of a milky way though its particular lighting. RGB linear LED lights were installed along its interior edges echoing with the glass cube.

The interior lighting also used different types and shapes of fixtures. While the optical fibers created the illusion of a starry sky, other functional lighting catched different patterns on the ceiling. Both the decorative and functional needs were met in the corridors, elevators and toilets.

The milky way under the glass skylight and the glass box under a dynamic changing colors, jointly form a magic treasure box, bringing energy and illusion. With such lighting design, the Jinyu Vanke Plaza is not mere a shopping center. Instead, it is an interesting space embracing recreation, entertainment and socialization.

第二章 商业综合体 | Chapter 2 Retail and Mixed Use
广州太阳新天地购物中心 | Guangzhou Happy Valley

照明初步概念图
Preliminary lighting concept rendering

名称：广州太阳新天地购物中心
地点：中国广州
完工时间：2012 年
总建筑面积：145 000 平方米
客户：广州华骏实业有限公司
室内设计：贝诺香港建筑设计有限公司
建筑设计：贝诺香港建筑设计有限公司
景观设计：贝诺香港建筑设计有限公司
照明设计：英国莱亭迪赛灯光设计合作者事务所 – 中国分部

Name: Guangzhou Happy Valley
Location: Guangzhou, China
Year of Completion: 2012
Gross Floor Area (GFA): 145,000 m²
Client: Guangzhou Huajun Industrial Co., Ltd.
Interior Design: Benoy (Hong Kong)
Architectural Design: Benoy (Hong Kong)
Landscape Design: Benoy (Hong Kong)
Lighting Design: Lighting Design Partnership International (LDPi) China Branch

照明设计的故事
Lighting Design Stories

惊喜！

广州太阳新天地购物中心位于广州市西黄浦路与马场路交汇处东南角的跑马场西边。由S303华南快速路出口驶向西黄浦路时，广州太阳新天地便可映入眼帘。贝诺所设计的建筑外形就像一个系有漂亮丝带的包装精美的宝盒，丝带环绕着建筑外墙，并最终在其上方形成了一个独特的可以随风水平摇摆的顶部结构。

Surprise!

Happy Valley is located on the west side of the former horse racing rink at the Southeast corner between the Machang Road and the West Huangpu Avenue. It is visible when one exits the Huanan Express way (S303) towards the West Huangpu Avenue.

第二章 商业综合体 | Chapter 2 Retail and Mixed Use
广州太阳新天地购物中心 | Guangzhou Happy Valley

最开始的灯光设计便是围绕着这个丝带展开,力争使其在夜晚也可一目了然。当然,随着建筑设计的进行,照明设计方案也在不断调整。最终,购物中心东西墙面上的丝带被大写的项目英文名字 HAPPY VALLEY 所取代。每个字母仅在底部施光,其框架结构则成为照明重点。为了使广告更加突出,建筑外墙其余部分并未布光。贝诺在购物中心内部设计了几处独具匠心的房顶,与走廊处简洁平整的房顶形成

The project façade designed by Benoy Hong Kong resembles a treasure box underneath a gift wrap with a beautiful ribbon tied around. This ribbon would stretch around frame structure forming the top part of the façade and would be swinging around its horizontal axis under the wind.
The original lighting design intent would have stretched this concept to

照明设计的故事
Lighting Design Stories

照明初步概念图
Preliminary lighting concept rendering

反差。尽管南北中庭在室内设计上有差别,但是设计精心的照明使二者协调统一。南中庭自底部到顶层呈椭圆状,扶梯顺时针旋转而上,于扶梯上仰望便可瞥见大气的椭圆形屋顶。扶梯底部以及屋顶处五个椭圆环的颜色与北中庭六至九层的颜色一致。北中庭顶部呈现出巨大的波浪延伸至南中庭,其两边巧妙地隐藏了 28 个 LED 射灯。新天地的室外景观中有多处很有意思的装置,均需要照明的配合。然而,最终实

make the entire ribbon visible at night. Nevertheless, the lighting design was changed as the shape of the ribbon was revised during the course of the project: it became a set of capital letters forming the name of the project in English ('HAPPY VALLEY') on the west and east facades, where each letter was highlighted from its bottom part while emphasis

第二章 商业综合体
广州太阳新天地购物中心
Chapter 2　Retail and Mixed Use
Guangzhou Happy Valley

施的方案中仅有一个装置进行了布光。东广场的小型LED地槽灯组成了12生肖的图案，在有效地吸引来往过客注意的同时，引导人们向着东入口行进。

在太阳新天地购物中心，宾客们不仅能有收获颇丰的购物体验，还能享受精心设计的照明所带来的连连惊喜。

was later set on the frame structure. The rest of the façade was intentionally left mostly in the dark in order for the advertisement to standout.

Inside the shopping center, Benoy's design includes a few interesting ceilings whereas the corridors' ceilings are clean and flat. Although the north and south atrium received a different interior design, they were

照明设计的故事
Lighting Design Stories

照明初步概念图
Preliminary lighting concept rendering

joined in terms of lighting.

The south atrium is oval in shape from the basement to the top floor with escalators twisting clockwise still providing a glimpse to the roof oval ceiling. The bottom of the escalators and the five oval ceiling rings were designed to have colors synchronized with the north atrium (floor 6 to 9)

照明设计的故事
Lighting Design Stories

which has a huge wave ceiling pattern all the way to the south atrium. This wave pattern ceiling was highlighted by 28LED projectors hidden on both sides.

For the landscape, Benoy's design included several interesting installations which required lighting. But ultimately only one has lighting: small ground recessed LED dots on the east plaza not only guide guests towards the east entrance, but also transform into the twelve zodiacal signs which appeal visitor's attention.

When guests come to Happy Valley, not only will they have excellent experience of shopping, but also they will be surprised by the lighting.

名称：北京来福士中心
地点：中国北京
完工时间：2008 年
占地面积：15 000 平方米
总建筑面积：110 996 平方米
客户：北京新捷房地产开发有限公司
室内设计：SPARK（原 SMC Alsop/Sparch Asia）
建筑设计：SPARK（原 SMC Alsop/Sparch Asia）
景观设计：SPARK（原 SMC Alsop/Sparch Asia）
照明设计：英国莱亭迪赛灯光设计合作者事务所 –中国分部

Name: Beijing CapitaLand Raffles City
Location: Beijing, China
Year of Completion: 2008
Site Area: 15,000 m²
Gross Floor Area (GFA): 110,996 m²
Client: Beijing Xinjie Real Estate Development Co., Ltd.
Interior Design: SPARK (formerly SMC Alsop/Sparch Asia)
Architectural Design: SPARK (formerly SMC Alsop/Sparch Asia)
Landscape Design: SPARK (formerly SMC Alsop/Sparch Asia)
Lighting Design: Lighting Design Partnership International (LDPi) China Branch

"会呼吸"的建筑

北京来福士中心位于北京核心商圈地带东直门西南角，由于地理位置的特殊性，设计师将大厦定义为时尚、充满活力的供年轻人休闲娱乐和购物的场所。在设计来福士广场的照明方案时，设计师并没有让它在原本已是灯火通明的东直门地带靠夺目的光效吸人眼球，而是选择在外立面适当铺设光源渐变的 LED 灯屏，夜

'Breathing' Building

Beijing Raffles City is located at the southwest corner of Dongzhimen, a core business area of the city. Given such special location, the positioning of Raffles City is to be a stylish and dynamic recreational venue and shopping center for the young. Instead of having dazzling lighting effects in an

北京来福士中心
Chapter 2 Retail and Mixed Use
Beijing CapitaLand Raffles City

晚远远看去，色彩的流转配合明暗度的转换，使整座大楼仿佛是"会呼吸"的建筑。照明不仅要体现建筑的透明结构，同时也需要利用不同的颜色表现外立面的丰富多样化，使灯光灵动、无眩光，同时也令人印象深刻。经实际测算，全部建筑外立面照明一整晚所耗电费低至395元人民币。

建筑外立面铺设的光源渐变的LED灯屏，主要源自于一个荷兰灯具生产商。由

already well illuminated area, the building's façade lighting design utilizes LED with gradient lighting. When night falls, looked from afar, the smooth change in colors, accompanied with the shift in brightness, gives the illusion that the building is breathing. Lighting is required to not only highlight the structure, but also bring changes on facades through different colors.

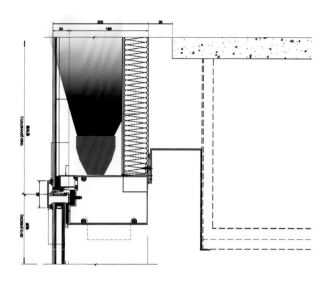

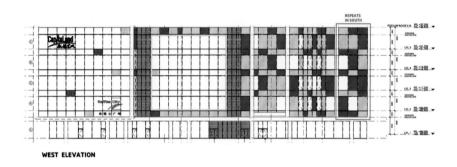
WEST ELEVATION

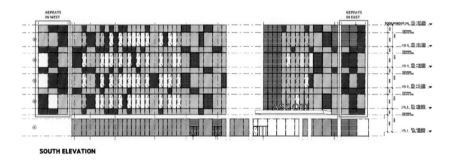
SOUTH ELEVATION

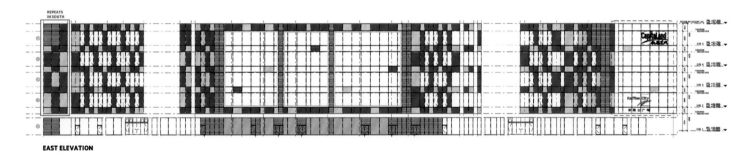
EAST ELEVATION

于外立面窗户的高度从 1.2 到 3.6 米不等，设计师运用了不同类型的照明灯具以满足不同高度的照明需求。

LED 灯具的动态控制和颜色变化可以通过 DMX512 进行预先设置，可以是均匀、单一、静态的颜色，也可以是单一的动态效果，或者是各种颜色轮番在立面上演各自的风采。灯具的选择是这个项目的难点，也是一个必要的过程，经过多次实

Meanwhile, it also needs to add dynamism to the architecture while eliminating glare and above all make the building more impressive. The cost of overall façade lighting per night is only RMB 395.00.

LED lights with gradient brightness and colors, installed on Raffles City's facades, are mainly from a Holland manufacturer. Since the heights of

验、评估和评级，所选灯具各个指标均达到了项目的要求：产品具有较好的性价比、良好的售后服务、较少的维护时间，且在业主的预算内；运行温度在 -20°—+50° 之间；能在特殊场合和节假日进行调光，以窗户为单位变幻出不同的颜色。建筑的外立面包括办公大楼和零售空间，要求灯具能够灵活安装且易于维护。尽管 LED 的寿命较长，但在安装的过程中不得不考虑其他的因素，如变压器、电线等，

windows range from 1.2 m to 3.6 m, different types of lights are applied to satisfy different needs.

Presetting DMX512 realizes the motion control and color changes of LEDs to achieve various lighting effects. It could be a static single-colored effect, or a dynamic single-colored one or even an effect with different

Lighting Design Stories

这就要求设备安装公司与照明设计公司相互配合,以确保灯具的安装正确无误,不会使某个窗户"鹤立鸡群"。Arup 设计单位提供了两套解决方案:零售店的立面灯具维护可以从橱窗内操作;办公大楼的立面灯具维护可以直接在户外进行。来福士的主要消费群体定位为 20 至 40 岁的时尚白领,因此其内部照明采用了最普遍的 T5 照明灯,以产生一种统一、明亮而又不刺眼的人工照明效果。与此同时,

colors showing off in turn on the facades. Selecting lighting fixtures is problematic in this project. They have to be cost effective, with great after-sale service, with less maintenance time and within budget; meanwhile they also need to be able to function between -20 ℃ and 50 ℃ and be adjustable according to various occasions and holidays. After

第二章 商业综合体 | Chapter 2 Retail and Mixed Use
北京来福士中心 | Beijing CapitaLand Raffles City

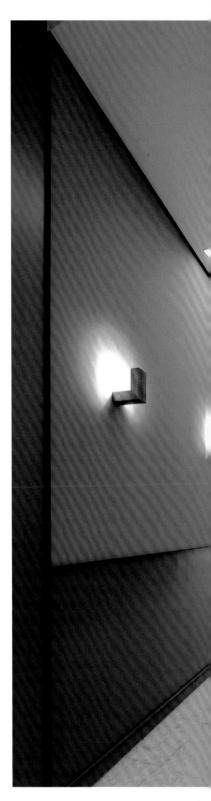

卫生间和走廊等处的照明细节也处处体现了人文关怀。起初，景观照明与内部照明通过安装在公共区域内地面和屋顶上的线型灯紧密相连。然而，由于质量问题，景观线型灯被形态各异的石头代替。

several examinations and assessments, every LED light chosen fulfilled the requirements in all those criteria.

The lights on the facades of the office and shopping center need to be flexible in terms of installation and easy to maintain. Despite the long operating life of LED lights, other factors such as transformers, wires and

Lighting Design Stories

etc. still need to be considered. Therefore, a sound cooperation with the facade lighting contractor was essential to ensure the overall accuracy in installations, avoiding any window standing out from the rest. The curtain wall company provided 2 maintenance solutions: LEDs of shops are maintained from the display windows and those of office buildings are maintained from outside.

The interior lighting uses the most commonly available T5 above a stretched material providing a uniform, bright and yet not glaring artificial lighting. Attention is also paid to the details of toilets and corridors, showing the care for humanity. Originally, both the landscape and interior lighting were connected through the forms of linear lights respectively installed in the ground and on the ceiling of public areas such as toilet, lift lobby... Unfortunately, the landscape linear lights were replaced with stones because of quality issues.

第二章 商业综合体 | Chapter 2 Retail and Mixed Use
成都凯丹广场 | Chengdu GTC Galleria

名称：成都凯丹广场
地点：中国成都
完工时间：2010 年
占地面积：20 800 平方米
总建筑面积：85 000 平方米
客户：成都环达通泰盈房地产开发有限公司
室内设计：Laguarda.Low Architects
建筑设计：Laguarda.Low Architects
景观设计：Laguarda.Low Architects
照明设计：英国莱亭迪赛灯光设计合作者事务所 –中国分部

Name: Chengdu GTC Galleria
Location: Chengdu, China
Year of Completion: 2010
Site Area: 20,800 m²
Gross Floor Area (GFA): 85,000 m²
Client: Chengdu Huandatong Taiying Real Estate Development Co., Ltd.
Interior Design: Laguarda.Low Architects
Architectural Design: Laguarda.Low Architects
Landscape Design: Laguarda.Low Architects
Lighting Design: Lighting Design Partnership International (LDPi) China Branch

照明设计的故事
Lighting Design Stories

蓝天之外

成都凯丹广场是 GTC 集团在中国的首个零售项目。广场位于机场高速与人民南路的交界处，正因如此，项目组也被赋予了一项极为重要的目标，那便是使凯丹广场达到 LEED (Leadership in Energy and Environmental Design) 金级。GTC 集团巧妙地邀请并聚齐了此项目经理下属的所有顾问，通过定期的电话会议和报告会，项目各方最大程度上保证了一致的进度和信息的共享。
由里及外，LLA 成功地使这个成都首家欧洲购物中心成为一个宾至如归的温馨的购物场所。

Out of a Blue Clear Sky

The GTC-Galleria Chengdu is GTC's first retail project in China. It is visible at the end of the airport express way before entering Remin South Road. As such it was given a very important goal for the project team to achieve: complete this project successfully with a LEED gold.
GTC made a very wise choice to gather all consultants involved under the leadership of its project manager. With periodic conference calls and report meetings, all parties were synchronized and in-line with each other.

第二章 商业综合体 | Chapter 2 Retail and Mixed Use
成都凯丹广场 | Chengdu GTC Galleria

建筑外墙冷暖元素交织，不同材料的运用以及被线型光源洗亮的各异造型加强了建筑整体的冷暖对比。

凯丹广场的天窗采用了特殊的照明处理手法，使得蓝天在夜晚仍能清晰可见。白天，零售区域运用弧线形的自然光，引导消费者找寻自己中意的品牌。广场内部从天窗到卫生间等许多地方采用了这种弧线形，配合灯光，形成了不同的阴影。

建筑外部设计以简单实用为宗旨，从而达到入夜时分行人、车辆在灯光的指引下进出流畅的效果。

然而，就在项目进入尾声时，一些元素却出乎意料地未达到项目组的预期效果。

For the entire project, from the outside to the inside, LLA made this first Euro shopping mall of Chengdu a very warm and welcoming place to come and shop.

The façade was designed to have clear separations between warm and cool elements based on differences in materials used, complemented by linear lights washing up the different patterns.

Several skylights received a special lighting treatment to have a blue sky at night. By day, they provide natural light into the retail area, with a curve

从照明的角度来说，以下几处仍需调整：首先，在资金允许的情况下，采用高质量 LED 灯具洗亮建筑外立面；其次，使用与屋顶灯箱透明度一致的材料做其他灯箱，从而保证将光源有效地隐藏起来；再次，在滚梯下方运用穹顶灯而非灯箱，不仅易于维护还可以节能；最后，为照明计划预留电路。

design allowing shoppers to look afar for their favorite brands. The curve pattern is also found in various locations inside the mall revealing different shades under lighting (from the skylights to the toilets).

The landscape was designed with simple functionality allowing a fluid entry and exit of both pedestrians and vehicles guided by lighting at night.

Nevertheless, towards the end of construction of this project, out of the blue, some elements turned out to be below the expectations of the project team. From the lighting point of view, the following would be the main elements to consider adjusting. First of all, use quality LED lighting fixtures to wall wash higher the façade, if budget allows; then use the same transparency for the material for light boxes as the ceiling light boxes to avoid seeing lamp sources inside; then use cove light under the escalators instead of light boxes to ease maintenance and reduce power consumption; at last, keep the circuitry available for the lighting schedules.

第二章 商业综合体 | Chapter 2 Retail and Mixed Use
北京冠城名敦道综合体 | Beijing Midtown

名称：北京冠城名敦道综合体
地点：中国北京
完工时间：2008 年
客户：鑫阳房地产开发有限公司
建筑设计：北京市建筑设计研究院、五合国际建筑设计公司
景观设计：AECOM Design + Planning / Landscape Architecture（原北京 EDAW）
照明设计：英国莱亭迪赛灯光设计合作者事务所 – 中国分部

Name: Beijing Midtown
Location: Beijing, China
Year of Completion: 2008
Client: Xinyang Real Estate Development Co., Ltd.
Architectural design: BIAD, 5+1 Werkhart International
Landscape design: AECOM Design + Planning / Landscape Architecture (formerly EDAW Beijing)
Lighting Design: Lighting Design Partnership International (LDPi) China Branch

照明初步概念图
Preliminary lighting concept rendering

从下到上

冠城名敦道综合体项目地处北京东二环和广渠门外大街两条城市主干道的交汇口。此项目是集商业、办公、居住、酒店于一体的多功能国际社区，标榜工作与生活体系合一，缩短生活需要的空间距离，提升都市生活质量的国际社区理念。

地产商采用的短期投资回报的商业模式使照明设计的经费非常有限。照明团队必须在所有的限制条件下仔细设计灯光，同时还要保证照明质量和租户与业主们在日常生活中对照明的使用。

在夜晚能够准确到位地展现出建筑的设计理念是件有意思但又很富有挑战性的

Bottom-up

Midtown is located at the intersection of the East 2nd Ring Road and Guangqumenwai Avenue, almost in the middle of everything. It is a multi-functional international community including the following functions: commercial, office, residential and hotel. The project aims to combine the work and life, simplify life and improve the quality of life.

The short-term return on investment business model of the real estate developer imposed a very strict cost control that only allowed a very lim-

北京冠城名敦道综合体
Beijing Midtown

事情。名敦道综合体的照明设计便是一个很典型的例子：此建筑为深色玻璃幕墙体结构，深色的材质吸光而玻璃的透光性又非常好，因此布光时遇到了些麻烦。设计师在建筑的办公和商业区域的外墙上尝试了几种投光方式，最后决定采用体积小并且易于隐藏的照明灯具，白天保持大楼的整体性，在夜晚为建筑带来韵律和生命。另外，照明设计还决定采用4 000 K白光与深色的建筑产生鲜明的对比。建筑外墙明框上施以主要照明，光源从下到上、由疏到密地变化排布，产生的节奏感照亮了主体建筑。设计师还在西北翼建筑的西北转角内侧辅以次级照明，用向上延伸的透光方式从内部清晰地照亮隐性幕墙的侧面，呈现出通

ited lighting budget. The team had to move forward by carefully designing lighting within those restrictions and at the same time keep the tenants and residents' perceptions towards lighting quality and what it brings to their daily life unaffected.

It is sometimes an interesting challenge to accurately express the concept of an architecture at night. And especially for this building complex: applying lighting on a large black stone or glass curtain wall is just problematic when considering that dark materials absorb lights while transparent glass

照明初步概念图
Preliminary lighting concept rendering

透感，从二环路上便可望见。

冠城名敦道生活居住区的照明亮度低于商业区，采用了简洁的 3 000 K 暖白光，与居住区建筑暖红色的外墙呼应协调。小区绿地和水景的照明则与景观主要路径两旁的功能性指引照明（庭院灯、矮柱灯、LED 地槽灯）合而为一。

lets light through if there is content behind or on it.

For the façade of the office and commercial areas, several ways of lighting have been considered and tried. But eventually the team carefully chose lighting fixtures that would have to be small enough to be unnoticed by day while bringing rhythm to the building by night. Next the team determined that the most suitable color temperature for this area would be 4,000 K to contrast with the darkness of the building. The team chose to install lighting in two locations: the primary lighting would be directly set on the

第二章 商业综合体 | Chapter 2 Retail and Mixed Use
北京冠城名敦道综合体 | Beijing Midtown

vertical fins of the façade, supported by a secondary unique vertical interior glow stripe on the northwest corner of the northwest building (noticeable from the 2nd Ring Road).

In contrast with the commercial area, the lighting of the residential area used a simple and comfortable 3,000 K warm white color temperature that matched the warm red of the residential façade, but certainly with decreased brightness. At the same time the lighting for the greeneries and waterscapes was merged with the functional guiding lighting used along the main landscape paths around residential compound (courtyard lights, pole lamps and LED ground recessed lights).

办公空间
Office

Chapter 第三章

03

北京凯德置地大厦（现北京 SK 大厦）
北京宝钢大厦
北京第五广场（现中国五矿大厦）
成都中海南城 1 号 B 地块 1 期
北京亮点设计中心
宁波市东部新城行政办公楼
哈尔滨哈西发展大厦
贵州花果园艺术中心
LDPi 北京办公室（2007—2012）

Beijing CapitaLand Tower (Now the Beijing SK Tower)
Beijing Baosteel Tower
Beijing Fifth Square (Now the China Minmetal Plaza)
Chengdu China Overseas Property South City Lot1B Phase 1
Beijing Liang Dian Design Center (LDDC)
Ningbo Eastern New Town Administration Zone
Harbin Haxi Development Building
Guizhou Huaguoyuan Art Center
LDPi Beijing Office (2007-2012)

名称：北京凯德置地大厦（现北京SK大厦）
地点：中国北京
完工时间：2006年
客户：凯德置地（中国）投资有限公司
照明设计：英国莱亭迪赛灯光设计合作者事务所－中国分部

Name: Beijing CapitaLand Tower (Now the Beijing SK Tower)
Location: Beijing, China
Year of Completion: 2006
Client: CapitaLand (China) Investment Co., Ltd.
Lighting Designer: Lighting Design Partnership International (LDPi) China Branch

白与金

凯德置地（中国）大厦坐落在东长安街国贸桥西南角，共37层，建筑总高140米，建筑外立面以石材与玻璃幕墙为主体。由于大厦是客户新加坡嘉德置地集团在中国的子公司凯德置地（中国）后期收购的项目，所以当LDPi团队加入时，其建筑结构和外立面都已施工完毕，这给照明设计造成了很大的限制。一般来说，照明设计师应在建筑设计完成初期加入到建筑、室内及其他专业设计团队中，这样可以最大限度地使得照明在整体上与建筑、室内设计形成浑然一体的效果，同时有利于项目各环节的同步协调、避免返工、降低成本。

在照明设计之初，客户希望不仅能够凸显建筑挺拔俊朗的风格，还要求能突出企业蓬勃发展、专业务实的内在精神。在与客户沟通后，设计师采用窄角度（3度角）2 000 W-4 000 K白色投光灯来表现建筑的挺拔。由于在建筑立面无法找到适当的灯具安装位置，设计师决定采用定制灯架来固定灯具。在设计灯架的过程中，特别注意了灯架的尺度、外观及结构。甚至设计了供灯具在现场安装时可以灵活调整位置的螺丝开孔。在用料及颜色等各个细节方面，力求达到与周围环境协调一致。灯具安装在同样的支撑架上，由于其投射角度的不同，起到了互相补光、明暗均匀的效果。使用窄角度灯具则使得所有的光照都能准确地投射在立面上，不会出现在空中散射和浪费的现象。为了营造蓬勃发展的企业形象，设计团队选择在建筑中间的U型槽和东西对称的立面用金色投光灯洗亮，增加建筑的活力和现代感，并在灯具外圈定制了金色滤色片来营造金色光效。对于企业标志周围的照明，则故意设计了阴影，使得蓝色的企业标志在夜空下很清晰。

凯德置地（中国）大厦周围景观呈现简约风格，绿植集中在大厦入口东侧，西侧则配以喷泉及装饰景观灯具。设计师针对主道路两边较为高大的绿植采用上照光对植物进行亮化，增加了树叶的自然表现力，其余树木则用4 000 K埋地灯打亮。大厦入口的西侧喷泉水体的自然喷出高度均在2米以下，照明不宜太过炫耀，选择由下向上的投射照明方式，结合周围的光环境自然的照亮，营造安静平和的氛围。设计师特意在建筑主体上用白色和金色布光，周边景观低调处理，凸显绿植、喷泉水景等自然的元素。夜幕下，月光、人造光、建筑、景观浑然一体。

White and Gold

CapitaLand Tower is located in the southwest corner of Guomao Overpass on the East Chang' An Avenue. It is 37-storey high, 140 m tall, built with stone and glass curtain wall as the main materials in its facades. Since this building was acquired by CapitaLand, when LDPi was engaged in the project, the building was almost completed, which brought significant limitations in the lighting design. Generally, lighting designers need to be involved in a project at an earlier stage when the architectural design is completed, together with architects, interior designers and other related professional teams. Therefore, the lighting would be able to be merged naturally with the building and interior design to a large degree; and meanwhile enhance cooperation and efficiency, avoid any reconstruction and reduce costs.

At the beginning of the lighting design, CapitaLand required not only to showcase the tallness and straightness of the building, but also reveal the development, professionalism and pragmaticism of the corporation. Based on the communication with the client, the 2,000 W-4,000 K, narrow-angled projectors were used to nail the straightness of the building. Since there were no appropriate locations identified on the facades for the installation of projectors, brackets were specifically made to hold them. Special attentions were paid to the measurements, shapes and structure of those brackets, even screw holes were designed to flexibly adjust the position of lights during the installation and maintenance. The details such as materials and colors were also carefully selected for a homogenous overall effect. The different projecting angles of the lights fixed on the same brackets supplemented each other, achieving a balance between brightness and darkness. The use of the narrow beams made it possible

第三章 办公空间
北京凯德置地大厦（现北京SK大厦）

Chapter 3 Office
Beijing CapitaLand Tower (Now the Beijing SK Tower)

照明设计的故事
Lighting Design Stories

that all are projected to the facade accurately, avoiding any light spill and enhancing the lighting efficiency. In order to create a brand image of a vigorously developing company, 'gold' light was projected in the central U-shape part of the façade while 4,000 K lighting was projected symmetrically on the east and west sides. In this way, the architecture looks more energetic and contemporary. The surroundings of the logo were intentionally left in dark so as to make the blue company logo stand out at night.

Simplicity is the foremost feature of the landscape surrounding the building. Plants were mainly located on the east side of the main entrance, while a fountain and decorative landscape lighting were installed on the west. The plants by the sides of the main pathway were treated with brighter lights coming from atop, thus emphasizing their natural beauty, whereas the rest of the trees were also highlighted by 4,000 K ground recessed lighting. Simple underwater up-lights were installed in the fountain where the height of each waterspout could reach 2m. With no exaggerations, the artificial light and natural light jointly create a peaceful and quiet ambiance.

The white and gold lights were selectively used on the main body of the building while simple lighting was used on the landscape to show its natural beauty of plants and fountains. When night falls, moonlight, artificial lightings, building and the landscape became a symbiotic whole.

第三章 办公空间 | Chapter 3 Office
北京宝钢大厦 | Beijing Baosteel Tower

名称：北京宝钢大厦	**Name:** Beijing Baosteel Tower
地点：中国北京	**Location:** Beijing, China
完工时间：2009 年	**Year of Completion:** 2009
占地面积：50 000 平方米	**Site Area:** 50,000 m²
照明设计：英国莱亭迪赛灯光设计合作者事务所－中国分部	**Lighting Design:** Lighting Design Partnership International (LDPi) China Branch

照明初步概念图
Preliminary lighting concept rendering

色温

宝钢大厦位于东长安街，是一座玻璃幕墙与钢构线条完美结合的、具有力度感和挺拔感的现代建筑。从外形上看，它彰显了钢铁工业带来的人类工业文明的崛起，也不失迈向现代主义的精神风貌，而如何将这种原始的美与现代意义的美相结合，成为建筑手段以外的重要课题，也为照明设计提出了不大不小的难点。

一座建筑在夜间的面貌必须成为设计过程中的一个要素，宝钢大厦是幕墙结构，由玻璃幕墙和铝合金构成，玻璃的反射透光，钢结构狭窄高耸的线条，都不是最好的承光载体。同时宝钢大厦毗邻东长安街沿线的银泰大厦、SK大厦、财源国际中心等一批近乎同时期建成的现代风格建筑，在高度上也都相差无几，如何能让它在这些建筑物中凸显出来，且能在夜色中融入到长安街的整体环境中，则更显得难上加难。

做照明的人都知道有个词叫"色温"。色温是用来形容光的冷暖的词语。色温低，即说明光源的颜色暖；色温高，说明光源的颜色冷。这就是民间所说的暖光、冷光。实际上，色温的概念是一位名叫Kelvin的英国人发明的。光源发射光的颜色与黑体在某一温度下辐射光色相同时，黑体的温度称为该光源的色温。在黑体辐射中，随着温度升高，光的颜色各不相同，黑体呈现由红—橙红—黄—黄白—白—蓝白的渐变过程。炼铁炉中的黑铁就是这样一个绝好的黑体。当温度逐渐上升至2 700 ℃时，铁会发出温暖的红色；而达到6 000 ℃时，炉中的铁则会呈现出淡蓝色。

当业主听到这个故事时，大楼的夜景方案即变得清楚：建筑底部入口的色温是温暖的3 000 K，中部立面采用4 000 K，顶部则是呈现出蓝色的6 000 K。整个设计不仅代表色温的攀升，更体现了作为中国最大、最现代化的钢铁联合企业——宝钢的企业特征。

同时，我们没有将建筑整体作为表现对象，而是将重点放在了建筑最具特征的部位进行局部特征的凸显：首先对建筑的重点部位进行强化；其次通过余光和环境光将整体建筑的序列性和元素的重复性体现出来，这样的表达绝没有半点多余的

Color Temperature

Situated on East Chang'an Avenue in Beijing, Beijing Baosteel Tower is a modern construction perfectly combining glass curtain walls with steel structure. Filled with strength, it points highly towards the sky. In its appearance, not only does the building show the rise of industrial civilization brought by steel industry, but also represents a positive spirit towards modernization. Yet, the challenge of the lighting design in this project lies in the harmonization of the building's sense of history with contemporary attributes.

The look of a construction at night is an essential part in lighting design. Constructed with reflective glass curtain wall and upright steep aluminum alloy structure, the facades of this architecture are not the best option for light. Neighboring with Yintai Center, SK Tower, Beijing IFC and many other modern constructions with similar height, it is challenging for Beijing Baosteel Tower to outstand from the crowds of architecture and integrate well with the surroundings of East Chang'an Avenue.

Color temperature is not a stranger to lighting designers. It is an expression to describe the warmness and coolness of lights. A low color temperature indicates a warm light source and vice versa. In other words, it is the so called luminescence and illuminine. Developed by a British named Kelvin, the color temperature is defined as the temperature under which the color of light radiated by a black-body is the same with that of a light source. In the black-body radiation, the color of light, as temperature rises, varies from red, orange, yellow, yellowish-white, pure white to bluish-white. The black iron in an iron-fining furnace is an excellent black-body. Iron becomes red when its temperature reaches 2,700 ℃ and turns into light

第三章 办公空间 | Chapter 3 Office
北京宝钢大厦 | Beijing Baosteel Tower

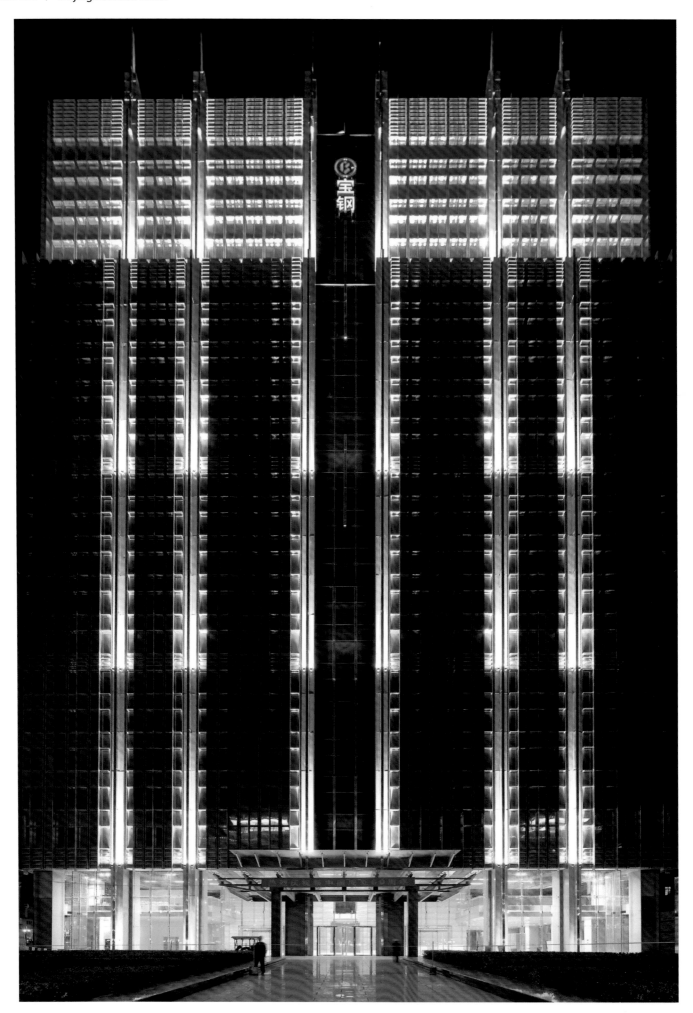

照明初步概念图
Preliminary lighting concept rendering

元素，又有很好的整体展示效果。

通过实景，我们可以充分感受到宝钢大厦建筑立面的通透感和向上延展的挺拔感，光带中采用阶段性的照度变化以产生节奏感，摆脱灯光单一性带来的简单呆板的感受，又有节节攀升的寓意与企业精神的传达。如果说照明手法本身就有叙事功能，那么顶部的宝蓝色配光正是与宝钢的企业标志色相吻合。这样的设计以精巧表现宏大，以简单传达出最大的企业自信，设计师认为这才是宝钢大厦建筑照明设计的最终目标。

blue when the temperature reaches 6,000 ºC.

The nightscape lighting design thus came into being inspired by the example of iron. The color temperature of the building entrance is set to be 3,000 K (warm), the part in the middle to be 4,000 K and the top to be 6,000 K radiating blue light. This is not only a design that grasps the essence of Kelvin's color temperature scale, but also a design represents the attributes of Baosteel, the largest and the most modern iron and steel enterprise in China.

Meanwhile, instead of focusing on the building as a whole, specific attentions are paid to the most iconic sections to make it more eye-catching. Not only the key parts are emphasized, but also the building's sequence and repetition of architectural elements are highlighted via residual light and ambient light. With no redundancy, the wholeness is perfectly interpreted.

The photos here show how neat and upright of the facades of Beijing Baosteel Tower. Light bands are designed to embody periodical changes in brightness, avoiding any dullness. Meanwhile, it also conveys the meaning of constant development and Baosteel's organizational culture. The blue light on the top corresponds with the corporate logo. This simple but well-designed lighting shows the grand scale of this tower and at the same time interprets Baosteel's confidence, which is believed to be the essence of the lighting design in this case.

第三章 办公空间
北京第五广场（现中国五矿大厦） | Chapter 3 Office
Beijing Fifth Square (Now the China Minmetal Plaza)

名称：北京第五广场（现中国五矿大厦）
地点：中国北京
完工时间：2007 年
总建筑面积：120 000 平方米
建筑设计：ARQUITECTONICA
照明设计：英国莱亭迪赛灯光设计合作者事务所 - 中国分部

Name: Beijing Fifth Square (Now the China Minmetal Plaza)
Location: Beijing, China
Year of Completion: 2007
Gross Floor Area (GFA): 120,000 m²
Architectural Design: ARQUITECTONICA
Lighting Design: Lighting Design Partnership International (LDPi) China Branch

三合一

第五广场（现中国五矿大厦）位于北京东二环东四十条西南角，分别坐落于北、西、南三个方位的 A 座、B 座和 C 座三栋写字楼通过大堂相连。

五矿大厦具有很特别的几何形状。从平面图上看，它们呈梯形，中央大堂和天窗为椭圆形。每栋楼都有一个独特颜色的三角形角：蓝色的 A 座的东南角，绿色的

Three-in-One

Beijing Fifth Square (Now The China Minmetal Plaza) is located at the southwest corner of the Dongsi Shitiao on the East 2nd Ring Road of Beijing. It has three buildings interconnected through a central lobby. Tower A, B and C are respectively the north, west and south building.

第三章 办公空间
北京第五广场（现中国五矿大厦）

Chapter 3 Office
Beijing Fifth Square (Now the China Minmetal Plaza)

B 座东北角以及黄色的 C 座东北角。

这独一无二的三角形成为每栋楼最引人瞩目的元素，也为照明设计团队带来了灵感，运用鹤来代表长寿。借鉴千纸鹤的理念，设计师们提出每个三角形的区域要各不相同，不仅勾勒出三角的线条，还要配合玻璃幕墙的颜色设计其最终的颜色。这些通过将玻璃内侧装的幕布用光打亮，突出每个角的颜色来进行实现。

室内照明在一定程度上与建筑外立面相连。每栋楼的椭圆形天窗与大堂入口相连，因此天窗上安装了与三角区域色彩相呼应但色调更加柔和的穹顶灯，营造一种轻松愉悦的氛围。

照明设计项目的进行过程中，由于 LED 价格依旧昂贵，建筑外立面、幕墙以及椭圆形天窗采用冷阴极灯。2012 年 8 月 6 日，随着中国五矿集团总部的正式入驻，三栋建筑外立面的照明色彩融合成为单一颜色，形成三合一的黄色角。

This project has specific geometrical forms. From a face and site plan point of view, they are trapeziform. The central lobby and its skylights are oval shaped. Each building has a unique colorful triangular corner: for tower A, B and C respectively it is blue on the southeast, green on the northeast and yellow on the northeast corner.

When people look at each building the most striking element is this unique triangular corner, which reminded the lighting design team of a crane, symbolizing immortality. Based on the concept of an origami crane, the team then proposed that each should be different, not only by outlining each corner, but also by filling each with a color based on the color of the glass curtain. For this purpose, LDPi proposed to light up the curtains installed behind the glass to strengthen the color effect on each corner.

The interior lighting is also to an extent, connected with that of the façade.

Lighting Design Stories

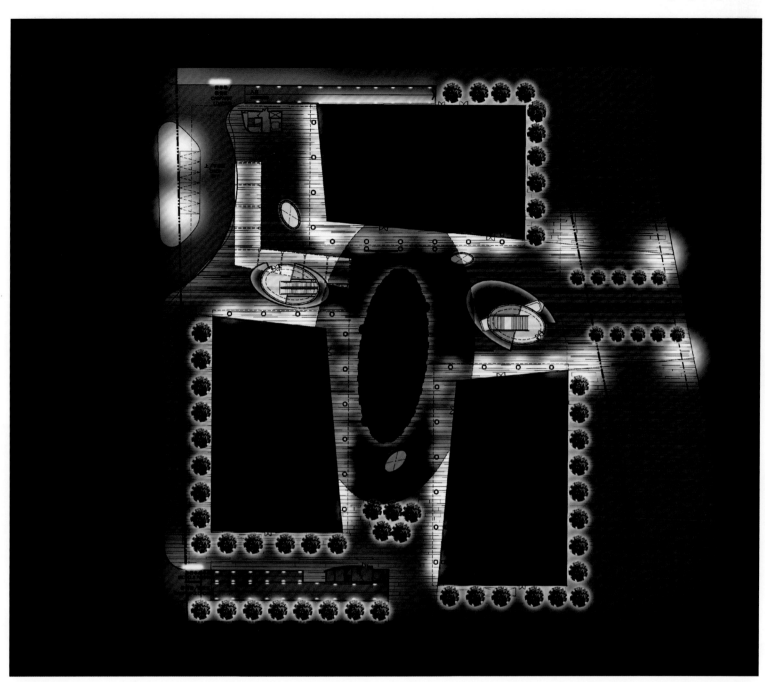

照明初步概念图
Preliminary lighting concept rendering

Indeed, each oval skylight is associated with the lobby entrance of a particular building. As such, each skylight was treated with a cove light with a color that matches that of the corner of its respective building, but only with a softer pastel tone to make visitors at ease.

At the time of the lighting design, because LED was still expensive, the customer had only enough budget to install cold cathode for the façade, curtains and oval skylights.

After China Minmetals' headquarter moved in on August 16, 2012 (thus renaming the building), the color of the façade lighting merged into a single color, making it a three-in-one yellow color outlined corner façade.

第三章 办公空间
成都中海南城1号B地块1期

Chapter 3 Office
Chengdu China Overseas Property South City Lot 1B Phase 1

名称：成都中海南城1号B地块1期
地点：中国成都
完工时间：2011年
总建筑面积：135 063平方米
客户：中海振兴（成都）物业发展有限公司
建筑设计：WONG & OUYANG(HK)Ltd.
照明设计：英国莱亭迪赛灯光设计合作者事务所－中国分部

Name: Chengdu China Overseas Property South City Lot 1B Phase 1
Location: Chengdu, China
Year of Completion: 2011
Gross Floor Area (GFA): 135,063 m²
Client: China Overseas Zhenxing (Chengdu) Property Development Co., Ltd.
Architectural Design: WONG & OUYANG(HK)Ltd.
Lighting Design: Lighting Design Partnership International (LDPi) China Branch

翱翔

中海地产强势进军成都，在新金融区1号B地块一期大范围兴建楼群。
此项目的外墙照明设计源于对建筑特征的准确理解以及对中海地产企业文化的把握。建筑设计以"双生"为理念，并在三处着重凸显：首先，两栋大楼一模一样，俨然是一对双生子；其次，双子塔在裙楼和上部有两处相连；最后，建筑的另一大特点是西楼西北角和东楼东北角相交成一造型奇特的转角，此转角并非简简单单地相交呈90度，而是组成了一个"M"，正是这造型特别的转角让照明设计团队开始从自然和富有历史感的元素中找寻灵感。
"M"形转角好似孩童稚嫩的画作中展翅高飞的鸟儿，翱翔于湛蓝的天空，沐浴着阳光，与白云为伴，俯瞰着大海。那么到底什么鸟儿才能契合中海地产的企业文化？海鸥，只有忠于同伴忠于领地的海鸥才是企业文化的最好体现。
由于海鸥多为白色或灰色，头部和翅膀上有黑色纹路，团队最终选择采用白色作为照明设计的主色彩，黑色则作为配色。设计还应用了照明调控系统，来帮助体现海鸥冲向高空时的力量和生命力，接着灯光逐渐变暗，黑色纹路则会慢慢显现。当成都中海地产南城1号B地块的所有项目完成时，它就像海鸥一样高高地在天空翱翔。

Soaring

China Overseas Property has landed in the new financial district of Chengdu. There will stand a set of buildings marking its strong presence point on lot 1B Phase 1 like animals naturally mark territories.

For this building façade lighting project, an understanding of the building specificities and inspiration from its corporate culture were at the source of the lighting design. Architecturally, the twin concept for this project is noticeable in three parts. Obviously the two buildings are mirrored and thus are twin towers. Then there is a twin connection between each other, one at the podium level and the other at the top. And the last unique feature is a special shaped corner located at the northwest corner of the west building and the northeast corner of the east building. Each corner is not just a simple 90 degree intersection between two lines. Instead, it is a corner in the form of an "M" letter. It is this special twin corner that started the team to look into natural and historical elements that would link lighting to this project.

That "M" shaped corner was like a child's drawing containing soaring birds high in the sky above the sea with a shiny sun, white clouds and blue sky. Which bird could be linked with the owner's corporate culture? The seagull it was. The seagull represents a dual faith (one of the corporate spirits of China Overseas Property): being faithful towards its mate and its site.

As a seagull generally is a white or grey bird with black marks on its head and wings, the team selected white as the main color used in the lighting concept. Nevertheless, black was still part of the concept but as a complement to the white through lighting controlling. Lighting controlling was introduced to help represent the strength and dynamism of the seagull's ascent and gradually the lighting dimmed to make the black wingtip visible.

Once all phases of the entire Chengdu China Overseas Property South City lot 1B are completed, it will be like a soaring seagull flying high above the rest.

Chapter 3 Office
Beijing Liang Dian Design Center (LDDC)

Name: Beijing Liang Dian Design Center (LDDC)
Location: Beijing, China
Year of Completion: 2012
Gross Floor Area (GFA): 9,000 m²
Client: Liang Dian Design Center (LDDC)
Architectural Design: MASAA
Lighting Design: Lighting Design Partnership International (LDPi) China Branch

北京亮点设计中心
Beijing Liang Dian Design Center (LDDC)

经验教训

自亮点设计中心改造项目之初,照明便成为项目组的核心。中心外部三层楼高具备展厅功能的"玻璃盒子"配以独到的纯色照明,是整栋建筑的一大特色,远远的从东四十条环岛便可看见。在改造过程中新添的另一亮点便是具有设计感的屋顶。另外,散布却引人入胜的点状埋地灯引领着访客经由便道走向中心。

Lessons Learnt

From the time of the renovation of the Liang Dian Design Center (LDDC), lighting was an integral part of the project team. From an exterior point of view, one of the key features of the building, with distinctive and clean color lighting treatment, is its 3-storey glass box serving as an exhibition

照明初步概念图
Preliminary lighting concept rendering

进入亮点中心，参观者有机会感受到照明的不同功能：玻璃展厅的顶端、电梯以及楼梯间都装以进口灯具进行装饰；在卫生间中，镜子后面安装了LED环境照明使得空间洋溢着温暖；展厅中，重点照明与参观路径紧密集合；公共空间采用荧光灯具作为功能性照明。
随着项目的不断推进，设计师总结出一些经验教训，在未来的项目可以引以为戒。

area on the ground floor. Besides the glass box being noticed from the distance of the Dongsi roundabout, the other element highlighted is the building's rooftop. In addition, discrete attractive lighting from ground recessed LED was designed to guide guests from the walkway into the building.

第三章 办公空间 北京亮点设计中心
Chapter 3 Office | Beijing Liang Dian Design Center (LDDC)

首先，增强信息流通。公制和英制衡量由于标准不同会引发问题，欧洲照明生产商与北美控制器制造商之间由于缺乏交流导致问题的产生。事实证明，将所有灯具混搭并按正相调光的顺序排列将造成高昂的花费。北美的控制器系统并不能调节从欧洲购买的照明器具。可行的解决方案是修改控制器系统订单，当然，这又会产生另一笔费用。

Entering the building, visitors have the chance to experience and feel different lighting functions: Imported decorative lightings are installed on the ceiling of the glass box, in the elevator and staircase lobby. Ambient LED lighting is used behind the mirrors to provide a warm feeling in the toilets. Accent lightings are used in combination with tracks in the exhibition area.

照明设计的故事
Lighting Design Stories

其次,将休假纳入施工考虑范畴。炎炎夏日,人人都应享有一定的调休时间。但这也影响了亮点设计中心的完工和开幕。由于工厂放暑假,自欧洲运抵的照明器材因为配件尚未到位而不能及时安装。因此需要等到工厂复工再进行安装调配。参与此项目的一位照明设计师表示,虽然改造过程压力重重,但是细细想来他为自己坚持到底毫无退缩而感到高兴。他在这个改造项目中受益匪浅。

Functional lightings are used in the public areas in the form of fluorescent lighting.

During the course of this project, there are few lessons learnt that can prevent pitfalls in subsequent projects.

Firstly, improve information flow: similar to problems arising from differenc-

Chapter 3 Office
Beijing Liang Dian Design Center (LDDC)

es between metric and imperial systems, problems arose due to the lack of communication between a European lighting manufacturer and a North American control system manufacturer. The common assumption that all lighting elements mixed and to be ordered as forward phase dimming proved to be costly. The result would have been that the North American

照明设计的故事
Lighting Design Stories

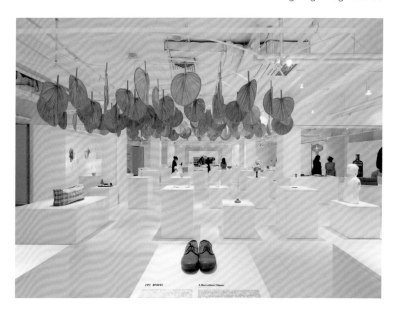

control system, being built at that time, would not have been able to dim the already purchased lighting fixtures from Europe. The viable solution was to adjust the purchase order of the control system, but of course with an additional cost for this change.

Secondly, consider holidays: everyone certainly deserves time off during

第三章 办公空间 | Chapter 3 Office
北京亮点设计中心 | Beijing Liang Dian Design Center (LDDC)

照明设计的故事
Lighting Design Stories

the summer. But that also proved to be impacting the site opening. Some of the lighting fixtures ordered were imported from Europe. While the fixtures arrived and were ready to be installed, some of the pairing accessories were still missing. The solution was only to wait for factory workers to return from their summer holidays.

One of the lighting designers who was onsite bared the construction pressure. Looking back, he was happy that he hadn't left the project. Instead, he held on and learnt quite a lot from this project.

193

宁波市东部新城行政办公楼
Chapter 3 Office — Ningbo Eastern New Town Administration Zone

名称：宁波市东部新城行政办公楼
地点：中国宁波
完工时间：2012 年
总建筑面积：271 898 平方米
景观面积：110 000 平方米
客户：宁波开投置业有限公司
建筑设计：上海秉仁建筑事务所
景观设计：SWA
照明设计：英国莱亭迪赛灯光设计合作者事务所－中国分部

Name: Ningbo Eastern New Town Administration Zone
Location: Ningbo, China
Year of Completion: 2012
Gross Floor Area (GFA): 271,898 m²
Landscape GFA: 110,000 m²
Client: Ningbo Development & Investment Real Estate Co., Ltd.
Architectural Design: DDB International Ltd. Shanghai
Landscape Design: SWA
Lighting Design: Lighting Design Partnership International (LDPi) China Branch

保守与现代

宁波东部新城行政办公楼位于新城核心开发区生态走廊以西，河清路与惊驾路东南交汇点。东部新城区作为宁波市未来的政治、经济、文化和商业中心与宁波老城区形成了"一市两核心"的格局。

此办公行政建筑群整体造型设计不同于以往衙门式政府办公建筑，而是呈钟形曲线。其南部宽广开放的广场与其北部高密度的建筑群形成"南松北紧"的鲜明对比。政府办公区的建筑群落包含一栋高为 90 米的主行政楼及六栋配套的附属楼，其中 D1-1a 为会议中心，D1-1b 为行政中心，D1-2a 为商务中心，D1-2b 为办公楼，D1-3a 为投诉及武警大楼，D1-3b 为后勤部。

政府建筑群另一特点是将以 D1-2b 为中心的网络变为横纵交错的平行线，从而为其他几栋楼的造型设计奠定了基础。这样一来各个建筑立面造型稍有不同，但建

Conservatively Contemporary

The seven buildings of the Ningbo Eastern New Town (ENT) Administration Zone are located to the west of the Eco-Corridor, at the southeast intersection of Heqing Road and Jingjia Road. Together with the old city center, ENT will form a 'one city, two hearts' structure with ENT being Ningbo's future political, economic, cultural and commercial center.

ENT's Administration Zone is different from traditional government buildings: the site is not restricted by any enclosure. The site is designed as a curve bell, with a fairly flat and huge open landscape square on the south part while the north part of the site is densely occupied by the site's

第三章 办公空间 | Chapter 3 Office
宁波市东部新城行政办公楼 | Ningbo Eastern New Town Administration Zone

筑肌理和属性表达与主建筑统一有序,并且整体通透、可视性强。

灯光设计的意图有多种多样,而这个项目的照明设计不仅满足了客户对照明最基本的需求,运用多数政府办公大楼都会采用的金卤射灯将建筑打亮,同时还融合了节能的理念,配以合理的调控,减少电耗和维修成本。在 3 000 K 暖白色灯光的铺陈下,政府楼群的韵律层次在夜间展现得淋漓尽致。在内庭、出入口、拐角等地方,稍稍增加了亮度来呈现整体建筑的体量和满足功能性照明。而后勤服务中心楼多用于会议和休息,在保持照明风格统一下特意减少布光从而降低亮度。

景观照明设计囊括了北部的建筑群中心和南部的广场。设计突出了几个区域的和谐,同时也强调了各区域的特质。景观照明不仅是景观设计的补充,还在能效的基础上有机地将景观和建筑统一起来。沿用了建筑照明的理念,也只用了单一的暖白光,没有多余的色彩和装饰。LED 水下灯打亮喷泉,内部园林通道采用矮柱

tallest building at 90 meters and six annex buildings (D1-1a conference center, D1-1b administrative center, D1-2a business center, D1-2b other offices, D1-3a complaints and armed police center and D1-3b logistics center).

Another architectural distinction comes from its conceptual design where a central grid (building D1-2b) is transposed by subtraction into vertical and horizontal parallel lines forming the pattern of the other buildings. Therefore, even though the buildings differ in terms of façade design, they are united to the core with the central building.

Although there were other lighting design intents, the team chose not only to match the requirements of the customer by simply highlighting

灯照明的同时打造安静的氛围。SWA设计的大面积起伏的带有植被的景观土丘成为外景的一大亮点,LED线型灯使其在夜晚看起来平铺的广场有了韵律和活力,又不失整体的安静与内敛。另一特色就是在广场多处选用了造型独特而简单的8米多高L型的高杆节能灯,用于功能照明的同时,也装饰美化着环境,呈现出一个现代化的广场。

照明设计除了考虑呈现完美的光效,使政府建筑更加亲民,还从客户的角度出发,遵从低调保守的原则。若这个项目是个商业建筑,则照明团队将会毫无疑问地突出网状格局和建筑外墙的横纵造型,并在景观照明上面采用艺术感强的照明装置。

the buildings in the same manner as most government buildings by using metal halide projectors, but also to integrate the principles of energy saving with a proper control to reduce power consumption and cost of maintenance. This time, these government buildings feel warm and welcoming under the 3,000 K color temperature and with a fresh new rhythmic look by seeing the combination of grids, horizontal and vertical patterns. Some areas were selected to be slightly brighter (inner courtyard, exits and corners) while the logistic center was intentionally designed to be dimmer than other buildings to avoid any disturbance as it is mainly used for conferences and recreational purposes.

The landscape lighting design includes not only the landscape of ENT's

第三章 办公空间
宁波市东部新城行政办公楼

Chapter 3 Office
Ningbo Eastern New Town Administration Zone

Administration Zone on the north, but also a central and a south landscape plazas. It emphasizes on the integration of these three plazas while preserving a variety of contents, especially water and landscape wave patterns. The landscape lighting not only complements the landscape design but also unites façade lighting and landscape lighting while energy efficiency being considered. The 3,000 K warm white color temperature used over the façade is also found in the LED underwater lights in the fountains, and bollard lights illuminates pathways in the garden, thus creating a quiet ambiance. One of the unique features of the landscape is the presence of mounds meticulously set on the plaza by SWA where linear LED lights modestly and quietly bring the undulations of the land-

scape into life. Another particularity of the plaza comes from the 8-meter-tall L-shaped LED energy-saving street lights installed along major pathways in various parts of the square which help modernizing the entire site while enhancing the functional lighting.

While the lighting design respects conservative requirements, it complements those by integrating contemporary technologies to keep warmness and friendliness of a government site by day and night. If this project were designed as a commercial project, the team would certainly have proposed to highlight the grid, horizontal and vertical patterns of the façade, while considering bringing lighting art installation into the landscape.

Chapter 3 Office
Harbin Haxi Development Building

Name: Harbin Haxi Development Building
Location: Harbin, China
Year of Completion: 2011
Site Area: 20,000 m²
GFA above ground: 15,000 m²
Client: Harbin Haxi New District Real Estate Development Co., Ltd.
Architectural Design: ZNA Beijing
Lighting Design: Lighting Design Partnership International (LDPi) China Branch

旋律与音符

建筑空间的美感来自空间的秩序和表面肌理的和谐。在哈西发展大厦照明设计过程中,强调主题和视觉中心,运用抽象的形体、光线、戏剧化的空间安排,着重提取建筑语言中最具特点的地方进行照明强调,成为设计师思考的重点。

哈西发展大厦是政府办公建筑项目,建筑设计由 ZNA 完成,位于哈西大街与学府四道街交口处,地上建筑面积 17 500 平方米,地下建筑面积 5 500 平方米,总投资 1.2 亿元人民币。由于大厦的设计中采用了现代、环保、节能的设计理念,也让这个政府办公大楼成为哈尔滨地区唯一一座现代化 5A 智能办公楼。

Melody and Note

The beauty of an architectural space comes in its order and the harmonization of surface texture. The lighting design of Haxi Development Building is theme-prominent and visual centric, with special consideration of applying abstract shapes and lights, clever arrangement of space and the highlight of the most symbolic parts of the building.

Haxi Development Building, located at the intersection of Haxi Street and Xuefu 4th Street, covers building areas of 17,500 m² above the ground

哈尔滨哈西发展大厦 | Harbin Haxi Development Building

Chapter 3 Office

从建筑的性质和未来使用的功能角度出发,设计师对这个项目进行了创造性的思考,整个设计把增进人类社交放在重要位置,巨大的弧线在一楼展厅和室外广场之间建立了较长的交互界面,巨大的柱列被安排成椭圆形,并环绕在前广场上,构成半环抱的平面形式。建筑师利用规则的柱列打破厚重板的体块结构,塑造轻盈的、竖琴般的纵向序列,仿佛走过去随手一挥,一系列悠扬的弦乐便会回荡在整个广场上。这种建筑形式加大了外立面照明设计的难度:既不能使石柱过于厚重粗壮,又不能暗淡无光被内透的室内光线所淹没。设计师从建筑结构出发,首先结合建筑玻璃体的通透性,充分发挥内透光的作用,让室内光成为建筑外立面照明的组成部分,包括柱廊的主立面,

and 5,500 m² underground. With a total investment of RMB 120 million, this government project is designed by ZNA (Zeybekoglu Nayman Associates Inc), incorporating modern, environmental-friendly and energy-conserving features, which make it the only 5A intelligent office building in Harbin, Heilongjiang Province, China.

Considering the nature and function of this building, its design creatively prioritizes the encouragement of social interactions. A giant arc is shaped at the massive interface of the ground floor and the square outside.

Lighting Design Stories

也通过内透光实现了华丽的呈现;其次通过自下而上的侧投光,恰好将浑圆的柱体变得柔和,富于变化。

哈西发展大厦的夜间景观照明将看似无序的、不相干的元素进行关联和组织,呈现出强化、甚至重构建筑的特点。精致的照明设计让建筑的视觉感受进一步延伸至景观广场,通过对水池和水柱进行照明,呼应建筑的柱列结构,形成视觉融合。在建筑周边,设计师针对树种特点和树冠树叶的形态特征,精心挑选灯具并通过光束角度、光型和光强的组合作用,为坚实单纯的建筑纵向线条补充了灵活生动的生命元素。廊柱、喷泉、树木在灯光的统一调度下,构成了一种协奏关系。

Enormous columns, arranged ovally, stand around the square, formulating a half-surrounded surface. The longitudinal and harp-like sequence of those columns brings liveliness and rhythm to the heavy and dull structure. This particular style adds the difficulty in the façade lighting design. A right amount of lighting is required, which is dim enough to adjust the scale of columns visually while bright enough to fit perfectly with the light coming out from the inside. The transparent glass curtain walls of the building enable light transmittance. Thus, interior lighting becomes a part of façade

第三章 办公空间 | Chapter 3 Office
哈尔滨哈西发展大厦 | Harbin Haxi Development Building

整体建筑照明以冷色调为主,通过灯光的精彩设计带给人们高效干练、现代清新的感受。多样性的内部空间带来视觉冲击,一改惯常办公楼直来直往一条通道的空间模式,在建筑内制造出优雅且带有变化的空间形式,并通过半透性玻璃隔断和变色的环境装饰光效,使一向显得呆板的办公空间活化起来,与建筑外立面的风格保持了相当的吻合度。

lighting and also embellishes the colonnade. Side lighting from the bottom up softens the columns, adding variations to the whole structure.

The nightscape lighting, combining random and unrelated architectural elements, stresses or even reorganizes the features of Haxi Development Building. Well-designed lighting extends the public's visual perception to the landscape square. Under the proper lighting, the pool and water columns echo with the main column structure, providing a unified vision. In the surrounding area, based on the characteristics of trees, trunks and leaves, carefully selected lights and a wise integration of angle, shapes

照明设计的故事
Lighting Design Stories

照明初步概念图
Preliminary lighting concept rendering

and brightness of light beams are applied which add vitality to the simple vertical lines in the whole structure. A harmonious concerto is thus created among columns, the fountain and trees.

Cool colors dominate the overall lighting of this building, conveying the sense of efficiency and freshness. This design breaks the stereotype of a one-corridor office building, bringing an innovative look by diversifying indoor spaces. Meanwhile, aided with semi-transparent glass screens and decorative lighting effects, the interior of the building not only is filled with vigor, but also grasps the essence of its exterior design.

第三章 办公空间 | Chapter 3 Office
贵州花果园艺术中心 | Guizhou Huaguoyuan Art Center

照明初步概念图
Preliminary lighting concept rendering

名称：贵州花果园艺术中心
地点：中国贵州
完工时间：2014 年
客户：贵阳宏益房地产开发有限公司
照明设计：英国莱亭迪赛灯光设计合作者事务所－中国分部

Name: Guizhou Huaguoyuan Art Center
Location: Guizhou, China
Year of Completion: 2014
Client: Guiyang Hongyi Real Estate Development Co., Ltd.
Lighting Design: Lighting Design Partnership International (LDPi) China Branch

山中的王国

地处贵州贵阳市的花果园是一家极具规模的地产公司,经营范围包括住宅、商务楼、写字楼以及酒店。其总部计划迁入现在的销售中心对面的新址并将华丽变身为艺术中心。所有设施不仅满足 5A 级写字楼的需求,同时也拥有足够的私密空间用于艺术品收藏。

King of the Hill

Huaguoyuan in Guiyang, Guizhou is a big property developer with residential, commercial, office and hotel projects already built, being designed or planned. The Huaguoyuan's own headquarter is scheduled to move to another location, across from its current sales center, and can only be qualified as

第三章 办公空间 | Chapter 3 Office
贵州花果园艺术中心 | Guizhou Huaguoyuan Art Center

照明初步概念图
Preliminary lighting concept rendering

在这个项目中，为了使这个11层高的总部大楼有足够的空间可以建在梯形结构的山丘上，周围的环境被重新改造了。建成后的艺术中心内部不仅拥有宽广的功能区域，包括办公室、会议室、陈列室、餐厅、休闲娱乐室、图书馆、走廊等，还拥有同等面积的私人空间，舒适奢华程度绝不输给任何国际五星级酒店。

这座半私人的建筑仅凭其独特的欧式外形便无疑会从其他建筑中脱颖而出。落成投入使用后，花果园艺术中心站在山腰俯瞰着自己的王国，迎接并盛情款待尊贵

a 'grand' art center. Everything there is made not only to accommodate a 5A grade office space for its staff, but it can also sufficiently accommodate private spaces, including rooms for art.

For such project, Huaguoyuan reshaped the surrounding environment in order to make room for its new 11-storey high headquarter to be constructed atop of the unique stair-like structure dug into hills. The interior was

照明初步概念图
Preliminary lighting concept rendering

的宾客。
艺术中心的照明设计力求满足多种需求。下表中列出了由1层至6层不同空间的照明类型的比例。

designed with a profusion of functional spaces (offices, meeting rooms, display rooms, dining rooms, entertainment rooms, library, corridors…) and commensurate private space kept undisclosed for this article. The center can definitely outpace the luxurious amenities of any international 5-star hotel brand.

This semi-private building will stand out from the rest as being the only European-style façade of such proportion in this area. And therefore, once

Chapter 3 Office
Guizhou Huaguoyuan Art Center

Preliminary lighting concept rendering

不同空间照明类型比例	重点照明	氛围照明	装饰照明	功能照明
走廊 / 大堂	15%	5%	40%	40%
陈列室	80%	5%	10%	5%
休闲娱乐室	20%	65%	10%	5%
图书馆	80%	5%	5%	10%
会议室	2%	8%	10%	80%
办公室	2%	8%	10%	80%
私人空间	待定	待定	待定	待定

in use, it will be sitting on a hillside overlooking its kingdom, while catering privileges to a unique crowd.

As such below is a table describing the proportion of type of lighting used per type of area (from floor 1 to 6).

Proportion of type of lighting used per type of area	Accent	Ambient	Decorative	Functional
Corridors / Lobby	15%	5%	40%	40%
Display rooms	80%	5%	10%	5%
Entertainment rooms	20%	65%	10%	5%
Library	80%	5%	5%	10%
Meeting rooms	2%	8%	10%	80%
Offices	2%	8%	10%	80%
Private areas	To Be Determined	To Be Determined	To Be Determined	To Be Determined

第三章 办公空间 | Chapter 3 Office
LDPi 北京办公室（2007—2012） | LDPi Beijing Office (2007-2012)

名称：LDPi 北京办公室（2007—2012）
地点：中国北京
完工时间：2007 年
总建筑面积：350 平方米
客户：英国莱亭迪赛灯光设计合作者事务所 – 中国分部
照明设计：英国莱亭迪赛灯光设计合作者事务所 – 中国分部

Name: LDPi Beijing Office (2007-2012)
Location: Beijing, China
Year of Completion: 2007
Gross Floor Area (GFA): 350 m²
Client: Lighting Design Partnership International (LDPi) China Branch
Lighting Design: Lighting Design Partnership International (LDPi) China Branch

"你好,再见!"

在2002到2013年间,LDPi团队在北京有过五个工作室。其中,只有一间是由LDPi自己设计的。为了拥有自己的空间,员工不再需要在京城早晚高峰时段排队等电梯,LDPi足足花了半年的时间才最终寻觅到了合适的办公地点。大家搬进了坐落于东城区美术馆后街的新办公室,这个过去曾经是泵房的砖石建

'Hello, Goodbye!'

From 2002 until 2013, LDPi's team was hosted in 5 different offices in Beijing. Only one of those five was uniquely designed by and for LDPi. It took half year to find a new and fourth office location allowing LDPi to not only have its own space, but also to avoid queuing for the elevator during

第三章 办公空间 | Chapter 3 Office
LDPi 北京办公室（2007—2012） | LDPi Beijing Office (2007-2012)

筑给整个印刷工厂带来清新的空气，独立的两层小楼为照明测试以及灯具储存提供了足够的空间。

放眼望去，整个区域仿佛是一片红砖建筑的海洋。LDPi 的新办公室则成为周遭第一座白色建筑。独特的外墙搭配上别出心裁的照明，将 8 米高的东外墙洗亮。荷兰产的线型 RGB LED 完美地体现出了砖的质感，并将建筑打造成了"钢琴键盘"。

the morning and evening rush hours.

LDPi's staff said 'Hello' to its new office located at the backstreet of the Art Museum in the Dongcheng District. Formerly it was a rubble room, providing fresh air for a printing factory, the vision was there: an independent two-storey loft allowing light testing and with room for fixture storage.

外立面照明不仅采用了DMX控制器变换每天的灯光颜色，同时还采用了天然的计时器。照明自日落时分自动开启，晚上十点准时熄灭。

办公室内一层的照明设计所采用的功能型吊灯源自于一个奥地利灯具生产商。T5灯管产生的光一部分直接打在办公桌上，而另一部分则导向了白色顶棚，进而反射到桌面上。2008年5月12日，在北京奥运会开幕前，四川发生了地震。当时

While the site was full of red bricks, LDPi's office became the first white building of the campus. With its unique façade, the early vision was embellished with a unique lighting. Linear LED RGB lights from a Holland lighting manufacturer were used to wall wash the 8-meter-high east façade, thus showing the texture of the bricks and depth of a piano-like keyboard. The

第三章 办公空间
LDPi 北京办公室(2007—2012)

Chapter 3 Office
LDPi Beijing Office (2007-2012)

员工们看到那些歪斜摇摆的吊灯才知道发生了什么。其余的办公空间都采用了奥地利和瑞士产的亮度可调的立式荧光灯和荧光吊灯。这些灯具都具备高质量的镇流器，所以办公室自装修到现在，尚未换过一盏灯。

二层会议室除了配有与一层一样的吊灯外，还安装了灯箱。因此能够在保证会议桌有充足照明的同时，投影仪又能清晰地显示出演示文稿。卫生间则是唯一的色

façade lighting was managed not only by a DMX controller allowing a one-color per day scene, but also by a natural timer turning the lights on at sunset and turning them off at 22:00 sharp.

The interior lighting on the first floor was mainly a mix of functional pendant lights from an Austrian lighting manufacturer. The fluorescent T5 tubes

温低于4 000 K的空间。在寒冷的冬天,2 700 K的T5荧光灯营造出温暖的氛围。办公室另一独到之处便是拥有三个灯光测试区域。LDPi项目中所建议使用的灯具都会提前在这里进行测试,查看不同高度下的性能并对结果进行评估。

好景不长,2012年国庆假期前的一纸通知将这一切都改变了。因整个区域的管理层变动,所有租户须另谋办公地点。LDPi不得不遗憾地跟这里说再见,向精心

had not only direct lighting onto the office desks, but also indirect lighting towards the white ceiling, which reflected some of the light back onto the tables. Just before the Beijing Olympics, those tilted pendants 'told' the staff an earthquake just happened on Monday May 12, 2008 far away in Sichuan. Other office spaces had dimmable fluorescent free standing

第三章 办公空间 | Chapter 3 Office
LDPi 北京办公室（2007—2012） | LDPi Beijing Office (2007-2012)

设计的立面照明、灯光测试空间以及温暖的卫生间照明告别。

最终，LDPi 搬到了离这个旧址不远的东城区亮点设计中心四层。原来所有的灯具都被再次利用起来。LDPi 也新添了照明设备，依照民间小折纸游戏来妆点特色墙和前台。

lamps from Austrian and Swiss manufacturers or ceiling mounted fluorescent lights. Because of the quality of the ballasts used in those fixtures, up to now, LDPi has not changed any fluorescent lamps.

The meeting room on the second floor had both functional lights (same pendants as on the first floor) and a light box focusing enough light above

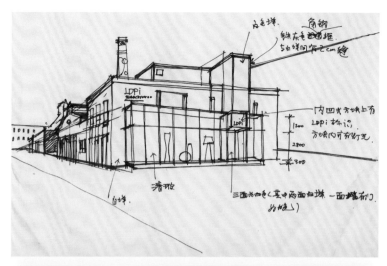

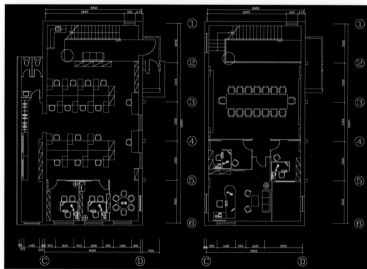

the meeting table while allowing a projector to show the content of a presentation. The only location which had a lower color temperature than 4,000 K was in the toilets, where a 2,700 K fluorescent T5 provided a virtual feeling of warmness during the cold winter (brick walls being not very insulating).

The other unique feature of the office was the three light testing frames allowing lighting fixtures to be tested at different heights, passing an evaluation before being recommended on LDPi's projects.

All of this suddenly changed before the October holidays of 2012: a notice from the campus indicated that it was going under a new management and that all current tenants would have to find new office locations. Therefore, LDPi had to unfortunately say goodbye to its façade lighting, light testing frames and toilet lighting.

Ultimately, LDPi stayed in the Dongcheng District and moved just a few minutes away from its former Beijing Art Museum location, saying 'Hello' to its new office on the 4th floor of the Liang Dian Design Center where all its previous lighting fixtures have been reused and a few new ones installed for a feature wall and front desk based on the concept of a cootie catcher.

住宅空间 Residence

Chapter 第四章

04

北京万科大都会住宅
北京红玺台住宅区
北京瑞士公寓
北京华远九都汇
北京丽都水岸

Beijing Vanke Metropolis
Beijing Eothen
Beijing Regent Land
Beijing Huayuan Jiuduhui
Beijing Richmond Park

名称：北京万科大都会住宅
地点：中国北京
完工时间：2013年
总建筑面积：42 000 平方米
客户：北京万科企业有限公司
室内设计：如恩设计研究室 (nhdro)
建筑设计：CABR
景观设计：如恩设计研究室 (nhdro)、北京麦田景观设计事务所
照明设计：英国莱亭迪赛灯光设计合作者事务所 -中国分部

Name: Beijing Vanke Metropolis
Location: Beijing, China
Year of Completion: 2013
Gross Floor Area (GFA): 42,000 m²
Client: Beijing Vanke Co., Ltd.
Interior Design: Neri & Hu Design and Research Office (nhdro)
Architectural Design: CABR
Landscape Design: Neri & Hu Design and Research Office (nhdro), MILAND DESIGN
Lighting Design: Lighting Design Partnership International (LDPI) China Branch

当灯光遇见奢华

最初接触北京万科大都会项目时，开发商多次强调"大都会"项目既不同于传统的豪宅，也不是传统的办公场所，"大都会"是CBD的私人会客厅，私享、社交的私人专属商务空间。内部设有大尺度的私享空间和丰富的功能空间，可按照个人喜好打造3D多媒体影院、名画收藏室、私人酒窖、雪茄屋、私人商务接待空间、

When Light Meets Luxury

At the beginning of the Beijing Vanke Metropolis project, the property developer emphasized that Metropolis, as an exclusive business space dedicated to private socializations, is different from traditional mansions and business venues. It has special private and multi-functional areas

第四章 办公空间 | Chapter 4 Residence
北京万科大都会住宅 | Beijing Vanke Metropolis

私人聚会空间、私人工作室等个性化需求空间，它将成为收藏、展示爱好的个人场所，也可以是私人接待、商务洽谈的场所。这是开发商对"大都会"定位的诉求，同时也是对设计师巨大的考验：如何让这样一个现代风格建筑在现代建筑云集的CBD商业中心脱颖而出，成为真正与众不同的"大都会"？

在进行设计之前，照明设计师对周边的光环境进行了分析：亮度偏高，光源混乱，

that could be built, according to personal preferences, into a 3D multimedia theater, painting collection room, private wine cellar, and cigar house, private business reception venue, meeting space, studio and many other individualized rooms. So, the Metropolis could either be a set of private areas for collections and exhibitions for personal interests, or exclusive

而"大都会"建筑的整体风格延续了标准的现代简洁造型,设计师最终把建筑外立面灯光设计主题定位于艺术、自然与品质。众所周知光是灵动的,每时每刻都在变化,并能改变它所触及的事物。建筑外立面的设计概念表现在瀑布一般的水流效果,设计师尝试使用人造光源来表现这种自然运动:用灯控软件设计控制不同的场景、不同的灯光的轮廓和灯光效果,使得外立面变得非常富有吸引力;采

places for businesses. Its distinct positioning challenges our lighting designers. In what ways could this stylish architecture truly stand out from all those modern constructions in the busy CBD?

Prior to the design, the analysis of the light environment of surrounding areas revealed the problems of over brightness and illuminant disorder.

第四章 办公空间 | Chapter 4 Residence
北京万科大都会住宅 | Beijing Vanke Metropolis

用经典的白色点光，映照出不同的画面组合；通过渐明渐暗的互动展示不同的韵律，这些巧妙的设计就像在夜晚给"大都会"披上一件奢华礼服，彰显着它的高贵与神秘。

入口处的灯光设计细致典雅，且覆盖了整个区域，营造出愉悦的氛围，并以此邀请参观者感受一种全新的生活体验。设计师将投射的灯光使用于墙体顶部，这种

So, to stress the contemporary and simple attributes of Metropolis, the mood of façade lighting is set to be artistic, natural and of high-quality. Light, so flexible, changes constantly and modifies surroundings. Artificial lights were applied in façade lighting to imitate streams of waterfalls, with software system controlling different scenes, shapes and effects, adding

效果使室内设计增添个性化元素，又抛弃了传统灯具的累赘感，尽可能地保持其高雅的特点。

对于"大都会"这个有着多功能、高品位的高端住宅，内部空间的照明理念秉承不喧宾夺主的主旨，遵循了室内设计专家的意见，将室内光线保持在中低照度，给人低调的奢华感，同时也让人感到放松，毕竟家是一个休息的场所。

the charm of the building. Classical white dotted lights projecting various scenes with interactions between darkness and brightness make the building even more luxurious. When night falls, Metropolis dresses up, showing its elegance and mystique to the public.

The lighting of the entrance is graceful and covers the whole area. A

北京万科大都会住宅
Chapter 4 Residence
Beijing Vanke Metropolis

室内功能空间的灯光设计是微妙的、抽象的、优雅的，灯具的外观简洁精致而又统一。灯光采用自动调节系统，允许使用者随意地增加或减少光的强度，而某些不同功能空间组合而成的场景效果中，灯光效果也将会是多种组合尝试的结合体。由此可见，一个好项目除了有好的建筑师和室内设计师，好的灯光设计师也至关重要，这些优秀的设计师能赋予这个建筑生命。

delightful ambience is created to encourage visitors for a brand new life experience. Spot lights, installed on the top of the walls, add personal features to the interior design while removing redundancy of traditional lamps to keep the elegancy of the area.

For this multifunctional high-end residence, the interior lighting design, fol-

lowing professional practices, is maintained at medium-lower brightness level to create a low-profiled and comfortable ambience. After all, home is a place for relaxation.

Lightings in functional areas are well-designed, blended with abstraction and elegancy. The appearances of lamps are simple, exquisite and unified. Their brightness could be automatically adjusted. Lighting effects in scenes jointly created by different multifunctional spaces are a mixture of various combinations.

Apart from outstanding architects and interior designers, professional lighting designers are also crucial to a real estate project. It is those talented designers that give the heart and soul of the architecture.

名称：北京红玺台住宅区
地点：中国北京
完工时间：2011年
占地面积：55 000 平方米
总建筑面积：130 000 平方米
客户：北京懋源置业有限公司
建筑设计：北京维拓时代建筑设计有限公司
景观设计：EDSA
照明设计：英国莱亭迪赛灯光设计合作者事务所－中国分部

Name: Beijing Eothen
Location: Beijing, China
Year of Completion: 2011
Site Area: 55,000 m²
Gross Floor Area (GFA): 130,000 m²
Client: Beijing Maoyuan Properties Co., Ltd.
Architectural Design: Beijing Victory Star Architectural & Civil Engineering Design Co., Ltd.
Landscape Design: EDSA
Lighting Design: Lighting Design Partnership International (LDPi) China Branch

融合

红玺台总占地面积约 55 000 平方米，总建筑面积约 13 万平方米，由 11 幢高层高端住宅建筑组成。它坐落于北京市东北三环外，与燕莎、CBD、亚奥等成熟商圈及通往首都国际机场的重要街港、北土城要道形成城市核心态势。它独具城市核心地珍奇资源：百年历史文脉、高自然景观资源占有率、高便捷度的城市动脉交通体系等。

中式的传统与现代在这个高端住宅区的建筑和院内景观设计中完美融合。灯光设计依循这一特色，在运用现代照明技巧凸显小区传统内涵的同时，还恰到好处地将小区与纷扰的外界隔离。考虑到住户休闲时光的私密性，照明巧妙地使整个小

Fusion

Eothen, covering 55,000 m² with a total building area of 130,000 m², is a residential block consisting of 11 high-rise buildings. Located outside the Northeast 3rd Ring Road, it forms the core of the city together with developed commercial areas such as Yansha, CBD, the Olympic Village and Asian Olympic Village Business Circle, arterial streets to the Beijing Capital Airport and the main streets in Beitucheng. This residential block has hundreds years of history, high occupancy of natural views as well as convenient transportation.

第四章 办公空间 | Chapter 4 Residence
北京红玺台住宅区 | Beijing Eothen

区在夜晚隐匿于京承高速旁。

红玺台的照明设计大体分为三个层次,从主干道望去,有着高辨识度的屋顶,邻近次干道的小区正门,以及建筑的入口和便道周围的景观元素。

寿喜纹作为装饰贯穿于整个小区的设计。楼顶、小区入口、建筑入口到院内景观,都能看到它的身影,灯光设计也自然不会忽略它。楼顶的前景照明、小区门口传统的红灯笼还有楼门入口的背式投光,无不着重强调着这个重要的装饰元素。

每栋建筑的楼号下面都配以巨大的青花瓷盘。这一设计不仅对建筑本身来说是极具巧思的,对于灯光设计来说亦是如此。其中一个瓷盘正面施以暖白色LED线型灯,而另一个屏风则是用LED从其背后做文章。

住宅区内景观设计采用王府园林"四进式院落布局"造园手法,以现代的表现形

This high-end residential compound is a fusion of traditional and contemporary Chinese lifestyle on both architectural and landscape sides. Following this theme, the lighting design's intent challenge was to reveal key elements with traditional meaning of this compound with contemporary techniques. At the same time, it must consider this inner area as a cutoff from the busy outer world. Private relaxing and resting personal time is also considered from the lighting point of view while the entire compound is invisible from the Jingcheng expressway at night.

The layers of the lighting design are split upon the rooftops visible from primary roads afar; the main compound entrance visible from the proxim-

式营造出东方园林移步换景的体验与自然之间的完美结合。
住宅建筑不仅照明层次分明，景观设计也是。针对小区入口处的石雕，设计师采用了聚光灯；在引导访客通往楼门入口的路径上则施以主要照明，而次要及再次照明则使用亮度逐渐递减的LED灯具。为防止眩光，地面安装了亮度较低的灯柱将光亮有效地引向地面。另外，绿植的照明设计采用传统上投光及防眩光控制部件的配合，来营造月光投影的效果。
水景设计也是红玺台住宅区的一大亮点。设计师仅采用了极简的照明，避免浮夸张扬。
在红玺台，优雅、传统和现代元素相辅相成同时又完美地结合，营造了一个和谐舒适的居住环境。在红玺台，如开发商最初所愿，人们真真正正地享受艺术人生。

ity of secondary roads; the buildings' entrances and landscape elements at the pedestrian level.
The entire site was designed with a key decorative element 'Shouxi' logo. It is found on the roof tops, at the main entrance, the buildings' entrance and in the landscape. Such an important element could not be missed by lighting, whether it is front lit on the roof tops, treated as a traditional red lantern at the main entrance or backlit at the buildings' entrance.
Another interesting architectural element is a huge Qinghua china disc below each building's entrance number. They are unique to each building in both content and lighting treatment. The disc is front lit with an LED

第四章 办公空间 | Chapter 4 Residence
北京红玺台住宅区 | Beijing Eothen

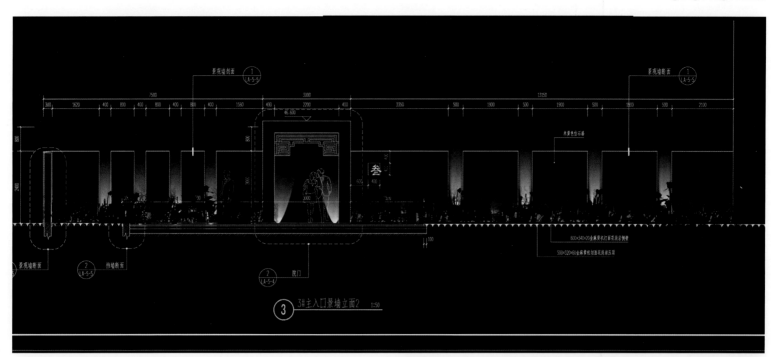

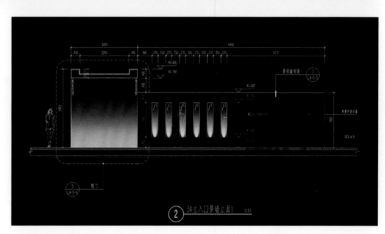

照明初步概念图
Preliminary lighting concept rendering

light on the outskirt of it while the folding screen is cleanly backlit with LED. The landscape design adopts the style of palace gardens. The block contains 4 courtyards and perfectly combines the oriental gardens with nature.

As the architectural buildings have layers of lighting, so does the landscape. From the main entrance, where a set of carved stones are the subject of a focal lighting, to areas leading guests to building entrances via key connecting pathways have a primary lighting level while secondary and tertiary become dimmer with LED fixtures. To avoid glaring, low light bollards were chosen to direct the light onto the ground. In addition to the traditional tree up-light, a few 'moon light' effect cast dynamic shadows with glaring control accessories.

Another significant element on the compound is the waterscape. As such it is simply treated by light but not profusively.

Elegant, traditional and contemporary elements echo each other and fuse with each other into harmonious and comfortable living environment. At Eothen, one really can fulfill and enjoy aesthetic life as the property developer intended to.

照明设计的故事
Lighting Design Stories

名称：北京瑞士公寓
地点：中国北京
完工时间：2008 年
占地面积：6 976 平方米
总建筑面积：40 358 平方米
客户：北京安泰置业房地产开发有限公司
建筑设计：深圳市三境建筑设计事务所
景观设计：北京建王建筑环境设计院有限公司（原北京建王技术咨询有限公司）
照明设计：英国莱亭迪赛灯光设计合作者事务所 –中国分部

Name: Beijing Regent Land
Location: Beijing, China
Year of Completion: 2008
Site Area: 6,976 m²
Gross Floor Area (GFA): 40,358 m²
Client: Beijing An Tai Real Estate Developer Co., Ltd.
Architectural design: G25 Architects & Associates (Shenzhen)
Landscape design: Jian Wang Design (Beijing)
Lighting Design: Lighting Design Partnership International (LDPi) China Branch

北京瑞士公寓 | Chapter 4 Residence
Beijing Regent Land

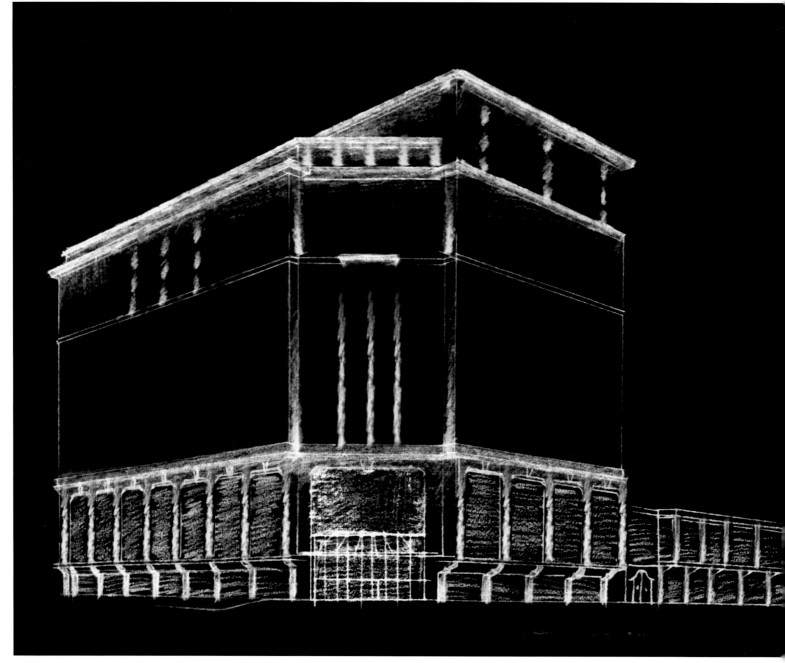

照明初步概念图
Preliminary lighting concept rendering

欧式经典

瑞士公寓地处北京港澳中心东侧，坐拥繁华的东二环商业区，采取商业与住宅相结合的设计，是真正高品质的公寓。

公寓建筑的外立面整体为新古典风格，西方古典建筑中大量的雕塑、图案和线角在这里被概括和简化为意向符号和细部装饰，这些元素增加了建筑的叙事性、经典性和鉴赏性。因此，在建筑照明手法上，设计师通过元素的选取和手法的构思，使光、形、色彩与环境氛围充分融合，与古典建筑风格意趣相投。

既要突出商业性，又要关注住宅的私密性和品质感，瑞士公寓的夜景照明设计首先设定主格调，遵循欧式古典的三段式建筑立面结构。

在腰线处，窗户檐口等是特别反映建筑语言符号的地方，设计师充分利用定向投光的洗墙光将浮雕的凹凸有致表现出来，同时又控制照度，使被照物不独立于整

European Classics

Located in downtown at the East 2nd Ring Road, to the east of Beijing Hong Kong Macau Center, Beijing Regent Land is a building combining both commercial and residential functions. It is a high-end serviced apartment.

The façade of the building is in neo-classical style. Sculptures, shapes, lines and angles in Western classical architecture are simplified here as symbols and detailed decorations, adding the narrative and ornamental value of the building itself. With careful selection of elements and ingenious techniques, the lighting design harmonizes well with the classic style, mixing brightness, shapes and colors together with ambience.

Lighting Design Stories

体，达到细节凸显的目的。顶部角线线条运用上投光，将檐口弧形斜面铺衬出面光，还顺势勾勒出一道清晰的亮线，形成了一种建筑的标志性符号。为了不影响住宅生活，多数上照的光线都采用光束极窄的灯具进行精准投光。

建筑不能孤立于环境，在景观照明方面，设计师有意打亮主路口行人步道，同时座椅的下方也安装了一些照明的灯具；其次，为了突出商业门面的商业氛围，只设置了一些必要的照明灯具与店招和灯箱进行相互补充；最后，在住宅的内庭院则设置了在保障安全前提下的最低照度值，形成安静私密的整体内部氛围。

瑞士公寓项目完工至今，已过去几年时间，从设计概念的产生到最后整体实现的过程，没有发生任何替换更改的现象，这一点也从侧面肯定了这个项目的成功，及委托人的认可。

To highlight the commercial attribute of the architecture while attending to the privacy and quality, the nightscape lighting of Beijing Regent Land sets the main theme and follows the classical style of three facades.

Windows and cornices at the building's waistline are the elements that often imply the symbolic language of the architecture. The outline of reliefs is highlighted by wash wall lighting; meanwhile the brightness is controlled. In this way, the illuminated objects are integrated perfectly with the entire building and the details are also emphasized. The spotlights that project to the corners and lines on the top of the building bring out the front lighting on the arc-shaped slopes of cornices, highlighting a clear line. An iconic symbol comes into being. The beams of most of up-lights are kept as narrow and precise as possible to minimize the impacts on the interior residential living areas.

Architectures could not be isolated from surroundings. For the purpose of landscape lighting, the main sidewalk is lightened up and some lighting fixtures are installed under the benches. Only necessary lighting is applied as a complement to store signs and light boxes, accentuating the commercial atmosphere. A quiet and private space is created with the lowest illuminance value but within safety standards.

Years have passed since the completion of this project, no replacements or modifications were made in this lighting design from the inception of concept design to realization of the design rationale. This project, being a success, gained the recognition of clients.

北京华远九都汇 | Beijing Huayuan Jiuduhui

第四章 办公空间 | Chapter 4 Residence

名称：北京华远九都汇
地点：中国北京
完工时间：2012 年
总建筑面积：108 756 平方米
客户：北京金秋莱太房地产开发有限公司
建筑设计：北京市建筑设计研究院
景观设计：EDSA Orient（北京）
照明设计：英国莱亭迪赛灯光设计合作者事务所－中国分部

Name: Beijing Huayuan Jiuduhui
Location: Beijing, China
Year of Completion: 2012
Gross Floor Area (GFA): 108,756 m²
Client: Beijing Jinqiu Laitai Real Estate Development Co., Ltd.
Architectural Design: Beijing Institute of Architectural Design
Landscape design: EDSA Orient (Beijing)
Lighting Design: Lighting Design Partnership International (LDPi) China Branch

照明设计的故事
Lighting Design Stories

见与不见

华远九都汇位于北京市朝阳区三环外霄云路上。东邻莱太花卉市场，南邻美国大使馆新址。

华远九都汇包含5座高端公寓、3座酒店大楼以及北区的商业区。1至3号公寓楼位于建筑群南侧可远眺朝阳公园。酒店大楼中，5、6号楼面对西北一带，7号楼面朝东方。此外，此三栋大楼的高度由北向南如金字塔状依次递增。

在LDPi于2008年初加入此项目时，华远九都汇的市场定位与现在的定位不同。起初，规划完毕住宅区、酒店以及相关设施的数量后，建筑群西侧的安排则是更注重于零售。然而，在2008年奥运会期间，开发商突然暂停华远九都汇这个项目并重新思考了它的市场定位。最终，此项目于2009年初华丽复工并成为如今成功的案例之一。

针对九都会的住宅区和酒店区，照明设计理念始终保持"见与不见"，不仅保证楼

See but Not Seen

The Huayuan Jiuduhui is located in the Chaoyang District of Beijing, right outside the 3rd Ring Road on Xiaoyun Road. On the east is the Laitai Flower Market and further south is the new campus of the US Embassy. The project is composed of 5 building mixing high-end apartments (building 1 to 3 on the south with a view on the north side of the Chaoyang Park when the weather allows), hotels (building 5 and 6 facing the northwest strip and 7 on the east) and with a bit of retail on the north point. Building 5, 6 and 7 have a special characteristic: from the north point on, they gradually rise up in a form of a pyramid towards the south residential buildings. When LDPi joined the project in early 2008, its position was slightly different. It was originally designed as project with a stronger retail focus on the

第四章 办公空间 | Chapter 4 Residence
北京华远九都汇 | Beijing Huayuan Jiuduhui

宇的私密性,同时照明灯光也保持低调隐约。不管是外墙还是景观,只可见柔和的光,但是光源均被最大限度地隐藏了起来。

建筑群北边的零售区域尽管规模有限,但是通过采用稍高亮度洗亮外墙从而使之脱颖而出。华远九都汇其他楼宇均采用柔和照明:6号楼的照明重点为入口处;1—3号楼则注重屋顶及内部入口的照明;5—7号楼的照明则侧重于高楼层,采用了壁灯和洗墙灯。

为了使景观照明简洁而人性化,设计师用光精致地区分不同的景观元素:矮柱灯被放置在交叉口;LED灯带被隐藏在台阶下;埋地灯打亮周围绿植;装有LED点光源灯打亮水景(有一些光源安装的位置比初始设计时略低);安装路灯符合安全标准。

事实证明,当初将华远九都汇重新定位的选择是正确的,也使得整个项目更为成功。

west strip while the number of residential and hotel building and am were settled.

Suddenly, in the middle of the 2008 Olympics, the customer put a h the project and took time to rethink about its positioning. The projec tifully resumed in early 2009 as the successful project known today For its residential and hotel areas, the concept of the lighting remained unchanged—'see but not seen'. This applies not only property itself as to keep the privacy to its occupants in such a lu environment and location, but also to the lighting. Indeed, either façade or in the landscape, one should only see the soft output of t but its source should not be seen to the best extent possible.

照明设计的故事
Lighting Design Stories

As the north point of the site, the retail, although limited in scale, was treated inversely by lighting to make it stand out from the rest of the project by wall washing the façade. For the rest of the buildings, they were treated with soft lighting: Building 6 focuses on its penthouse entrances; buildings 1 to 3 focus on the roof tops and the inner entrances; building 5 and 7 focus on the rising parts, respectively with wall lamps on the terraces and wall washing lights.

To match the simplicity and humans scale of the landscape, lighting was installed meticulously to mark the separation of landscape elements: Bollards at specific intersections, steps with LED stripes hidden underneath, few ground recessed lights to highlight a handful of trees, waterscape with sparkling LED dots (with a quantity installed lower than the design intent), pole lights to comply with safety standards.

Looking back, the choice to adjust the position was absolutely correct and made the project even more successful.

第四章 办公空间 | Chapter 4 Residence
北京丽都水岸 | **Beijing Richmond Park**

名称：北京丽都水岸
地点：中国北京
完工时间：2006 年
总建筑面积：435 000 平方米
客户：中国房地产开发基金会
建筑设计：Langdon Wilson International
景观设计：EDSA Orient（北京）
照明设计：英国莱亭迪赛灯光设计合作者事务所 - 中国分部

Name: Beijing Richmond Park
Location: Beijing, China
Year of Completion: 2006
Gross Floor Area (GFA): 435,000 m²
Client: China Real Estate Development Fund
Architectural Design: Langdon Wilson International
Landscape Design: EDSA Orient (Beijing)
Lighting Design: Lighting Design Partnership International (LDPi) China Branch

照明初步概念图
Preliminary lighting concept rendering

国画

丽都水岸位于北京市四元桥东南,朝阳酒仙桥路以西、坝河以北、芳园西路以东。建筑所处地理位置优越,公共设施齐备,交通便利,购物方便。其临河的一侧极为繁华。

丽都水岸按照新古典主义原则进行设计,并结合现代建筑中明快大气的建筑风格,体现出简洁、细腻、典雅的特点,整体设计融于自然。建筑在3、18、20层采用弧形设计将楼体聚拢。

通常,住宅小区的灯光设计在体现住宅楼体建筑特点的同时还要控制灯具安装的位置、数量、功率、亮度以及颜色,以确保居住者的正常生活不受影响,保证私密性,让居住者有安全、和谐、舒适感。设计师对建筑结构作了整体了解后,决定用简洁的光勾勒出建筑的特点。此外,18层弧形的灯光设计区别于3和20层以达到更好的视觉效果。

考虑到客户的预算,设计团队运用4 000 K冷阴极灯,对三个弧形建筑结构进行照明处理。在第一次现场试验后,设计师们发现实际效果细腻不足,生硬有余,18层的特点得不到体现,达不到设计预期。由于更换LED会大大增加成本,设计师在继续沿用所选定灯具的基础上,对线型冷阴极灯做了柔化效果处理。在18层的弧形上用黑色遮光罩遮光,利用遮光罩安装位置的不同形成一个波纹组,对弧形建筑结构进行柔化处理。这种渐变的效果使建筑在夜晚充满生机。

从坝河望去,丽都花园南侧起伏的立面与河道交相呼应。北侧立面则采用射灯凸显建筑的高耸和刚劲。

在2006年那就是最好的方案了,如果是现在我们重新设计,我们会考虑采用LED灯,在简化工程、方便维护的同时更好地控制LED光点的密度,实现更理想的照明效果。

夜色中,单色冷白光勾勒出的丽都水岸,与周边的河流、绿植等美景相映成趣,犹若一幅以光做墨绘制而成的中国传统国画。

Traditional Chinese Painting

Richmond Park is located southeast of the Siyuan Bridge, between the west of Jiuxianqiao Road, north of Ba River and east of Fangyuan West Road. Not only has it convenient access to several parks and public transportation and shopping, but also its surrounding is flourishing with its south elevation facing the Ba River.

Richmond Park, adopting neo-classicism principles and combining contemporary style, is simple, elegant, detail-oriented and fairly natural. It bears a unique architectural feature: a sinusoidal pattern wrapping the building all around its 3rd, 18th and 20th floors.

Generally, for residential projects, the lighting design must ensure that the lighting is controlled in terms of location, quantity, power, brightness and color in order to show the unique features of the building. At the same time, privacy of the residents must be kept to the largest extent so that one can enjoy a peaceful and comfortable rest without being disturbed by the façade's artificial lighting. After having completed a thorough study of the Richmond Park and its surrounding environment, the team decided to use simple and clean white light to highlight and express the characteristics of the architecture at night. The team also chose to bring a slight difference between the sinusoidal pattern on the 18th floor and the other two in order to make the building appealing to the eye.

Considering the budget constraints of the client, the team chose to use 4,000 K cold cathode within the three sinusoids. However, during the first lighting test, the team noticed that the outcome differed slightly from the design intent: the overall effect was too strong and uniform thus hiding details on the 18th floor and making the building less attractive. Instead of considering replacing the cold cathode with a costlier LED light source, the team recommended a cheaper solution by using a custom-made black louver installed on the top of the lighting fixtures along the sinusoid of the 18th floor in order to allow more or less light output based on the location of the louver on the sinusoid to create the wave pattern. With the louvers in place, details became softly highlighted while the curves were smoothly and gradually lit, bringing a flow of life to the building.

Standing across the Ba River, one could then see the entire wavy south façade fluidly and perfectly reflect on and merge with the course of the water. Furthermore, the north elevation received an additional treatment with projectors installed to showcase the height and strength of the building.

This entire installation was the most effective at the time of the design in 2006; however, if it were redesigned today, LED would be primarily considered for the convenience of the possibility to directly control the intensity of each LED dot and of the maintenance.

At night, Richmond Park forms a unique and timeless traditional Chinese painting by integrating its wavy artificial facade lighting with the natural surrounding greeneries and Ba River.

文化教育及其他
Cultural Space and Others

Chapter 第五章

05

北京万科如园
固安规划馆
上海长颈鹿美语学校
成都东站

Beijing Vanke Ruyuan
Gu'an Planning Exhibition Center
Shanghai Giraffe English Training Center
Chengdu East Railway Station

第五章 文化教育及其他 | Chapter 5 Cultural Space and Others
北京万科如园 | Beijing Vanke Ruyuan

名称：北京万科如园
地点：中国北京
完工时间：2011 年
总建筑面积：85 099 平方米
客户：北京五矿万科置业有限公司
建筑设计：北京市住宅建筑设计研究院有限公司
景观设计：北京清华城市规划设计研究院
照明设计：英国莱亭迪赛灯光设计合作者事务所 –中国分部

Name: Beijing Vanke Ruyuan
Location: Beijing, China
Year of Completion: 2011
Gross Floor Area (GFA): 85,099 m²
Client: Beijing China Minmetal Vanke Real Estate Co., Ltd.
Architectural Design: Beijing Residential Architectural Design Institute Co., Ltd.
Landscape Design: Beijing Tsinghua Urban Planning and Design Institute
Lighting Design: Lighting Design Partnership International (LDPi) China Branch

光与影

万科如园是中国五矿地产与万科合作的住宅项目,位于北京西北5环向北4.5公里外。如园地理位置十分优越,毗邻百望山森林公园、颐和园、圆明园。

小区的照明设计包括A单元样板间的室内照明、销售中心的外墙和外景照明以及园区内有300多年历史的永庆堂的外墙照明。永庆堂原本位于山西省,而后被拆卸并重建于北京。

每一位参观者从踏入如园开始要先后经过四个区域,见证自然景观与人工照明相

Light and shadow

Beijing Vanke Ruyuan is a joint residential project between China Minmetals Real Estate and Vanke. It is located in Beijing, 4.5 km north outside the Northwest 5th Ring Road. This site is in a very privileged area nearby the Baiwang Forest Mountain Park, the Summer Palace and Yuanmingyuan Imperial Garden (both on the southeast).

The lighting design scope of work for this project was the interior lighting

北京万科如园
Chapter 5 Cultural Space and Others | Beijing Vanke Ruyuan

结合所产生的光影世界,随着参观者逐渐深入到如园,光影世界的内容也愈发充实。如园的主入口处有一巨石横卧在门边,上书"如园"。入夜,一排小射灯将其打亮,简洁而大方。由此处,访客会到达主建筑边的停车场。在大门左边的墙边矗立着20根如古时栓马石似的景观装饰柱,上面雕刻着人和动物的图案。错落安置的埋地灯将这些柱子的质地凸显出来,在夜色中更加体现出空间的深度。

经由大门,访客路过地槽灯会看到两座波浪形的隔断砖墙,墙面由内而外被照亮,

of a mockup room for its unit A, the façade and landscape of its sales center; and the façade of a unique 300 year Yongqing Tang old house. Originally from the Shanxi province, this house was dismantled and rebuilt element by element in Beijing.

Each visitor would go over the site through four areas sequentially and witness a game of light and shadow created by natural landscape items

光线从其开口处一半的位置柔和地透出,留下的另外一半则在昏暗中只能看出轮廓。在这两座墙后,矗立着重建的永庆堂。两组地槽灯分别勾勒出建筑的外墙。
在永庆堂和隔断墙之间是一块方形的平面喷水池,上面有如园的地形图。喷水池被右侧墙上的灯光打亮,不过所购买到的灯具以及配件并未契合设计初衷。另外,方形喷水池的四边也都安装了线型的灯具。
暂别永庆堂,访客向售楼中心方向走去,会路过一处夸张的有着特色墙面的景观,

under artificial lighting. The game rises in terms of content as visitors' further move inward.

Starting from the main road entrance, there sits a huge horizontal stone where the site's name Ru Yuan is simply lit up by a row of small projectors at night. From there, visitors reach a parking area near the main site entrance gate. Beside the wall on the left of the main site entrance gate,

北京万科如园
Chapter 5　Cultural Space and Others
Beijing Vanke Ruyuan

由水、树和竹子构成。此处的特色墙与之前的隔断墙搭建理念如出一辙，亮度较高的砖面清晰可见，而另一边则隐藏在黑暗之中。

售楼中心高两层，周围有巨型柱子，灯光从侧边打来，使得中心的正面除了屋顶以外全部被黑暗笼罩。售楼中心边上有一个非常精致的竹园，仅有墙面被打亮，与内部漆黑一片的竹子形成鲜明对比。尽管如园的售楼中心面积有限，但景观环境使得它妙趣横生，配以含蓄的照明，似有微风拂过，光影婆娑。

there are twenty ancient like landscape stone carved pillars, Shuanma Zhuang stone, mixing human and animal figures. Its textures are captured by a handful of ground recessed lights randomly installed, thus creating a sense of depth together with darkness.

After walking through the main entrance gate guided through ground recessed fixtures, visitors meet two partition brick walls. Being lit from the

inner side, light softly and gradually escapes through half of its opening while the other more visible half remains obscure. These partition walls stand ahead of the reconstructed Yongqing Tang old house where two pairs of ground recessed lights near each of the outer columns highlight its façade.

In front of the old house and behind the partition walls, lay a square flat fountain acting as a map and introduction of Ru Yuan. Although it is tentatively lit by lighting installed on the top of the right partition wall, the purchased fixtures and its accessories did not match the design intent. The alternate linear lights softly wash the four side of the square.

After leaving the old house and moving towards the sales center to the right, visitors reach a profuse landscape with feature walls, waterscapes, trees and bamboos. This time, as feature walls are built in the same manner as the previous partition walls and are being seen from the inside, the brighter half of the bricks forms the most visible part while the other less visible half remains somber.

The sales center is a two-storey high building with magnificent columns lit from the side, leaving the front part in the dark while highlighting the roof part. On the side of the sales center rests a quiet small bamboo garden where the wall is lit up with contrasting black bamboos at night. Although the Ruyuan's sales center's site is relatively small, the landscape environment makes it interesting when playing with subtle lighting and counting on the wind to make oscillating shadows.

第五章 文化教育及其他 | Chapter 5 Cultural Space and Others
固安规划馆 | Gu'an Planning Exhibition Center

照明初步概念图
Preliminary lighting concept rendering

名称：固安规划馆
地点：中国固安
完工时间：2012年
占地面积：28 807.89 平方米
地上建筑面积：8 805 平方米
客户：固安幸福基业资产管理有限公司
建筑设计：Valode & Pistre（法国 VP）
景观设计：WSP
照明设计：英国莱亭迪赛灯光设计合作者事务所－中国分部

Name: Gu'an Planning Exhibition Center
Location: Gu'an, China
Year of Completion: 2012
Site Area: 28,807.89 m²
GFA above ground: 8,805 m²
Client: Gu'an Happiness Foundation Asset Management Ltd.
Architectural Design: Valode & Pistre (France VP)
Landscape Design: WSP
Lighting Design: Lighting Design Partnership International (LDPi) China Branch

Lighting Design Stories

2^4

固安展馆位于正在飞速发展的城市固安,毗邻计划于2015年落成的北京第二国际机场,建于大型水景之上,与一连桥相连,出入甚为方便。建筑由16个方块组成,包含展示区和办公区。每个方块的外周,规格为16米×16米,由大小为2米×2米的玻璃或者石头构成。

固安展馆利用展馆自身的几何形状来进行照明设计保证了建筑结构从各个角度都

2^4

This museum project is located in the rapidly developing city of Gu'an, nearby the planned second Beijing international airport due to open in 2015. Built atop of a large waterscape and accessible via a connecting bridge, the 16 cubes composing the pavilion include exhibition and offices spaces. The peripheral faces of each cube, measuring 16 m × 16 m, are

固安规划馆 | Gu'an Planning Exhibition Center

Chapter 5 Cultural Space and Others

清晰可见。

客户还有一个特别要求：入夜时分，即便展馆闭馆之后室内照明已关闭，但是飞机即将降落在北京第二国际机场时，旅客们依然能够看到建筑结构。每个方块间都巧妙地加入了照明设计，四个投光灯安装在建筑底部，灯带安装在底部和顶部，这一设计使得从空中望去屋顶就像四连战游戏棋的棋盘。

从街边望去，建筑四边清晰可见。入口处的发光的玻璃方块就像个灯笼，从此处

each subsequently composed of glass or stone, measuring 2 m × 2 m each.

From a lighting point of view, using the geometrical shapes of this museum ensured to make the structure visible from various point of views.

The client specifically requested that when flight would land at the second Beijing international airport, passengers would see the structure at night

Lighting Design Stories

开始,向建筑四角方向针对玻璃窗的重点照明逐渐减弱,方块内玻璃窗下方照明灯具的密度减小,照明效果开始递减。

行人能够围绕亭子散步,寻找更多有趣的景观。玻璃光盒的照明动静兼具,不仅好似静浮于水池之上,还好像划过星空的拖着星尘之尾的动态流星。置于水池底部的灯具将水纹映到博物馆外墙上,这一景象亦妙趣横生。

even if interior lighting is turned off after the museum's closing time. Indeed, dedicated lighting was added between cubes (four projectors at the bottom and linear lights both at the bottom and the top), making the roof look like the base of a Quarto board game from above.

At the street level, the four sides of the building would also become visible, starting from the main entrance with the glass box glowing as a lantern,

with a gradual fading effect based on the decrease of glass windows concentration towards each corner of the building as well as the decrease of intensity of the dedicated lighting installed inside beneath each glass windows.

As pedestrians have the opportunity to walk around the pavilion, other interesting views awaits them at night. It has the duality of being both static and dynamic: not only would this treasure glass box be stationary floating above the pond, but also it would be an animated reflection of shooting star with its trailing dust tail. The second view that wanderers could enjoy is a transition of the water coming from the pond onto the facade of the museum as wave patterns emerge from light source hidden below the surface of the pond.

上海长颈鹿美语学校
Shanghai Giraffe English Training Center

第五章 文化教育及其他 | Chapter 5 Cultural Space and Others

名称：上海长颈鹿美语学校
地点：中国上海
完工时间：2009 年
总建筑面积：2 000 平方米
客户：上海徐汇区长颈鹿美语培训中心
照明设计：英国莱亭迪赛灯光设计合作者事务所－中国分部

Name: Shanghai Giraffe English Training Center
Location: Shanghai, China
Year of Completion: 2009
Gross Floor Area (GFA): 2,000 m²
Client: Shanghai Xuhui District Giraffe English Training Center
Lighting Design: Lighting Design Partnership International (LDPi) China Branch

照明设计的故事
Lighting Design Stories

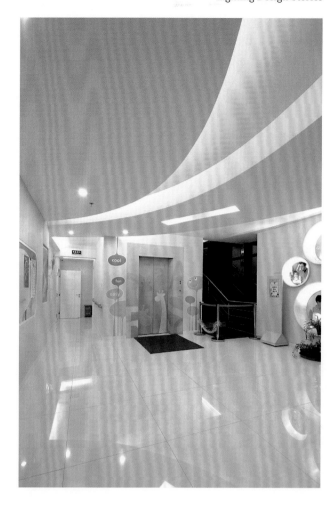

基石

将色彩法则运用于环境设计，从选取最佳视域，设计注意力的捕捉物，用光照的强弱和投影的有无来呈现目标的质感、结构、色彩甚至轻重，到安排视觉流向的诱导因素，规划视觉流程的形式，再到最后印象的形成，这一系列过程称为环境中的视觉引导。根据不同的使用目的，环境必须考虑色彩要求、性格体现、气氛形成，其中如果没有科学合理的用光，色彩的作用可能会被大大削弱。

照明设计则需要考虑几个因素：首先是环境功能对照度的要求；其次是用什么样的光色和照明方式才能呈现出材料的质感；最后建筑结构或装饰物的立体性的呈现。

Cornerstone

Visual guide in environmental design consists of the application of color scheme, selection of the best field of vision, the design of eye-catching objects, the brightness of lights and the existence of shadows showing the characteristics of the objects, to the arrangement of factors in visual flow, the planning of visual process, and finally to the formation of an image. According to different purposes, the design of environment should consider colors, the characteristics reflected and the overall ambiance. Without appropriate lightings, the effect of colors is likely to be significantly

上海长颈鹿美语学校
Chapter 5 Cultural Space and Others
Shanghai Giraffe English Training Center

长颈鹿学校的照明设计过程，在保证满足功能要求的基本原则下，充分地发挥了灯光对色彩的表现力。室内设计需要形成心理中心，而这个中心常常会与空间中心是重合。空间使用属性不同，条件不同，中心的处理手法也大相径庭。一些大型空间可以用一组装饰、花坛或雕塑构成视觉上的焦点，产生中心地位。而在这样的小型空间特别是儿童活动区，地面尽量开敞平坦，家具尽量紧凑低矮，空间的视觉中心就会自然向上发展，顶部处理成为关键。为了形成空间的中心位置，长颈鹿学校学生活动室顶部被设计成晴朗的天空，中心部位置一灯槽并配合放射性的线型灯槽。这一手法丰富了顶面的构成变化，恰似天空的太阳，照耀着下面幸福欢乐的孩子们，更可以理解为一朵绽放的花朵，成为这个空间的主角。其实

minimized.

Several factors need to be considered in lighting design, such as brightness according to the function of the venue, the color and way of lighting that could highlight the quality of different materials and the presentation of the three-dimensional features of a building or an ornament.

Upon meeting the functional requirements of the Giraffe English Training Center, the concept of lighting design has also expressed the colorful environment. A psychological center, necessary in interior design, is often coincident with the center of a space. Designing manner of a psychologi-

照明设计的故事
Lighting Design Stories

不管寓意如何，这里的照明设计绝不仅仅是简单的、充满诗意的，它包含着许多专业的设计和考量。

这个项目的照明设计对于环境中光的分布要求很高，既不能让孩子们在阅读的时候受到投影的影响，又不能让美丽的童话书在这里失去鲜艳的色彩，光源的选择和整体配光的计算非常关键，以形成均匀完整的空间光影关系。室内设计师只需要单一思考材料的安全性、色彩关系和材质的牢固度，但是没有灯光师的配合这一切可能都将在光的干扰下变了格调。

家长接待区的功能比较复杂，在整体风格和照明手法上，既要延续学校整体环境风格，又要考虑成人的视觉习惯和动作要求。在这里，家长可能在此静坐等候，

cal center varies hugely among different attributes and conditions. A set of decorations, a parterre or a sculpture can be the visual focus of a large space. Where as in a small space like this school, the ground area was kept wide and flat, equipped with compact and low-rise furniture. Hence the visual center shifts to the top, making the design of ceiling a key element to this project. The activity room is designed with a sunny ceiling with both a central cove light and radiating cove lights. This enriched ceiling is just like the sky, where the sun is shining vigorously over the children who play in the room. It can also be seen as a flower becoming the center

第五章 文化教育及其他 | Chapter 5 Cultural Space and Others
上海长颈鹿美语学校 | Shanghai Giraffe English Training Center

也可能会相互交流,还可能要阅读一些文字材料,因此照度的调试,要满足各种行为的需要,并呈现出最合理的氛围。

通过对长颈鹿学校整体环境的深入理解,精妙绝伦的灯光设计不仅满足了功能上的需求,同时还赋予这群天真烂漫的孩子们一个绚烂、趣味横生的环境。

of this particular space. In fact, the lighting design here is not just about seeking for simplicity and poetry; it also involves professional design and scientific measurements.

The lighting design of Giraffe English Training Center has high requirements on light distributions. The arrangement should minimize shadow while the children are reading and meanwhile maintain a colorful and clear environment. Hence, it is crucial to select lighting sources carefully and calculate overall distribution of lights for an even and complete relationship between lights and shadows. Interior designers only need to consider

Lighting Design Stories

2009

2010

2011

2012

2013

the safety of materials, colors scheme and solidity of materials. However, without the cooperation of lighting designers, the style of all those features may well be modified by lights.

The reception area for parents has complicated functions. Not only the style and means of lighting here should fit in with the overall design of the school, but adult's habits and behaviors also need to be considered. In this area, parents may sit and wait, chat with each other or even do some reading. Therefore, the brightness should meet the requirements of all those activities and at the same time help to create the most appropriate overall ambiance.

Through a better understanding of the environment, the elaborate lighting design of Giraffe not only meets functional demand, but also makes the colorful environment stand out with a sense of fun for the keen hearted children.

第五章 文化教育及其他 | Chapter 5 Cultural Space and Others
成都东站 | Chengdu East Railway Station

名称：成都东站
地点：中国成都
完工时间：2011年
总建筑面积：108 000 平方米
客户：中国中铁二院工程集团有限责任公司
室内设计：中国建筑西南设计研究院有限公司
建筑设计：中国中铁二院工程集团有限公司
照明设计：英国莱亭迪赛灯光设计合作者事务所－中国分部

Name: Chengdu East Railway Station
Location: Chengdu, China
Year of Completion: 2011
Gross Floor Area (GFA): 108,000 m²
Client: China Railway Eryuan Engineering Group Co., Ltd.
Interior Design: China Southwest Architectural Design and Research Institute Corp., Ltd.
Architectural Design: China Railway Eryuan Engineering Group Co., Ltd.
Lighting Design: Lighting Design Partnership International (LDPi) China Branch

白加黑

如今的中国各城市飞速发展,联系愈加紧密,新的火车站也迎来愈来愈多的高速铁路列车。几乎每座火车站都包含了具有当地文化特色的设计元素,成都东站也不例外。

成都东站占地 108 000 平方米,于 2011 年投入使用。车站内部分为 5 个区域,

White and Black

With the rapid development of Chinese cities and how they are interconnected, new train stations welcome the Chinese high speed train. Each station bares local cultural elements in its design and so does the Chengdu East Railway Station.

第五章 文化教育及其他 | Chapter 5 Cultural Space and Others
成都东站 | Chengdu East Railway Station

照明初步概念图
Preliminary lighting concept rendering

均可经由户外广场或者两架高架道路到达。站内设施齐全，包括高架候车大厅、售票处、位于中层楼的餐厅和购物区、14个站台和26个通道、出口（售票处）和两层成都地铁线路。

政府要求站内所有公共区域进行照明模拟，以确保功能性照明达到建筑照明设计标准 GB 50034—2004。这也意味着进行装饰性照明和塑型照明的空间非常小。

The Chengdu East Railway Station, put into use in 2011, has an area of 108,000 m² divided into five main interior levels accessible either from the outdoor Square or from two elevated roads: An elevated Waiting Hall (ticket offices, with mezzanine eating areas and shopping area); Platform (ground level with 14 platforms and 26 tracks); Arrivals (ticket offices); Two levels

唯有叶形顶棚的照明可以有发挥的空间。树叶的造型经过几次修改，因此其照明形式也有些变动，从最初的在树叶中脉进行布光，改变成在树叶边缘安装灯具。火车站外部的照明设计则包含了更多古老的蜀文化元素。从照明的角度来说，主入口、铜雕塑、波浪形的中层楼外立面、车站标志、檐口、石头基座、立柱以及下客区域，这八个部分需要夜景照明。其中，仅有下客区域需要照明模拟以确保

of Chengdu Metro lines.

For all of those interior public areas, lighting simulation was required to show officials that Standard for lighting design of buildings GB50034-2004 would be met for the functional lighting. This meant that there was little or no room for decorative or accent lighting. The only area that required

第五章 文化教育及其他 | Chapter 5 Cultural Space and Others
成都东站 | Chengdu East Railway Station

照明初步概念图
Preliminary lighting concept rendering

达到标准。

然而,由于经费限制,所有的外立面照明均未实施,仅有下客区域和车站标志采用了照明。夜幕下,火车站显得分外低调、不显眼。

尽管成都东站的室内外照明具有强烈的反差,然而二者的结合犹如钢琴键盘一样,黑白分明。

more in-depth design cooperation with the interior designer was on the special leaf shaped ceiling. As the design of the leaf was changed a few times, it also changed the lighting layout from being at the central midrib to finally being at the outer skirt of the leaf.

The exterior design features decorative elements of the ancient Shu cul-

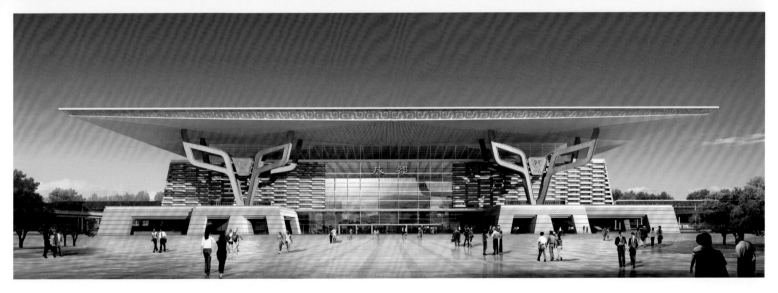

ture. From a lighting perspective, those are among the eight contents considered to be lit up by night (main entrance, bronze elements, mezzanine façades, logo, cornices, stone pedestals, side columns and drop off area). Here lighting simulation was only required at the drop off areas in order to ensure that standards are met.

第五章 文化教育及其他 | Chapter 5 Cultural Space and Others
成都东站 | Chengdu East Railway Station

照明初步概念图
Preliminary lighting concept rendering

Unfortunately, due to budget constraints, most of the exterior façade lighting was cancelled, leaving the train station unrecognizably black at night with only lighting for the drop off areas and logos.

Although there is a huge difference between the outcome of the interior and exterior lighting, both coexist forming a huge white and black contrast

as the keys of a piano keyboard.

英国莱亭迪赛灯光设计合作者事务所 – 中国分部
地址：中国北京市东城区东四十条 94 号亮点设计中心 412 室
电话：+86-10-52978528
邮箱：Enquiry@LDPinternational.com.cn
网址：www.LDPinternational.com.cn

Lighting Design Partnership International (LDPi) China Branch
Add: Liang Dian Design Center (Room 412), 94 Dongsi Shitiao, Dongcheng District, Beijing, P.R. China 100007
Tel: +86-10-52978528
Email: Enquiry@LDPinternational.com.cn
Web: www.LDPinternational.com.cn

图书在版编目（CIP）数据

照明设计的故事：汉英对照／英国莱亭迪赛灯光设计合作者事务所中国分部编著. —
天津：天津大学出版社，2014.7
ISBN 978-7-5618-5004-6

Ⅰ.①照… Ⅱ.①英… Ⅲ.①照明设计—中国—图集 Ⅳ.①TU113.6-64

中国版本图书馆CIP数据核字（2014）第145100号

《FRAME 国际中文版》杂志社 策划
地　　址：北京市朝阳区望京西路48号金隅国际E座12A05室
电　　话：010-84775690/5790
邮　　箱：info@designgroupchina.com
责任编辑：路建华　叶　玮
装帧设计：陈　强　师　岚
封面设计：陈　强

Lighting Design Stories
照明设计的故事
英国莱亭迪赛灯光设计合作者事务所—中国分部 编著

出版发行：天津大学出版社
出 版 人：杨欢
地　　址：天津市卫津路92号天津大学内（邮编：300072）
电　　话：发行部：022-27403647
网　　址：publish.tju.edu.cn
印　　刷：北京信彩瑞禾印刷厂
经　　销：全国各地新华书店
开　　本：230mm×300mm
印　　张：17.25
字　　数：302千
版　　次：2014年8月第1版
印　　次：2014年8月第1次
定　　价：248.00元